Mixing with Impact

In *Mixing with Impact: Learning to Make Musical Choices*, Wessel Oltheten discusses the creative and technical concepts behind making a mix. Whether you're a dance producer in your home studio, a live mixer in a club, or an engineer in a big studio, the mindset is largely the same.

The same goes for the questions you run into: where do you start? How do you deal with a context in which all the different parts affect each other? How do you avoid getting lost in technique? How do you direct your audience's attention? Why doesn't your mix sound as good as someone else's? How do you maintain your objectivity when you hear the same song a hundred times? How do your speakers affect your perception? What's the difference between one compressor and another?

Following a clear structure, this book covers these and many other questions, bringing you closer and closer to answering the most important question of all: how do you tell a story with sound?

Wessel Oltheten has been recording and mixing music since his teens, which has led to a successful studio business with a very diverse clientele. To hear some of his work you can visit **www.wesseloltheten.nl**. For extra articles and a video course accompanying this book, visit **www.mixingwithimpact.com**.

T0139328

Mixing with Impact

Learning to Make Musical Choices

Wessel Oltheten

Translated by Gijs van Osch

Routledge
Taylor & Francis Group

NEW YORK AND LONDON

First edition published 2018
by Routledge
711 Third Avenue, New York, NY 10017

and by Routledge
2 Park Square, Milton Park, Abingdon, Oxon, OX14 4RN

Routledge is an imprint of the Taylor & Francis Group, an informa business

© 2018 Wessel Oltheten

The right of Wessel Oltheten to be identified as author of this work
has been asserted by him in accordance with sections 77 and 78
of the Copyright, Designs and Patents Act 1988.

All rights reserved. No part of this book may be reprinted or
reproduced or utilised in any form or by any electronic, mechanical,
or other means, now known or hereafter invented, including photocopying
and recording, or in any information storage or retrieval system,
without permission in writing from the publishers.

Trademark notice: Product or corporate names may be trademarks
or registered trademarks, and are used only for identification
and explanation without intent to infringe.

Previously published in Dutch by Edusonic 2016

This edition published by Routledge 2018

Library of Congress Cataloging-in-Publication Data
Names: Oltheten, Wessel, author. | Osch, Gijs van, translator.
Title: Mixing with impact: learning to make musical choices /
written by Wessel Oltheten ; translated by Gijs van Osch.
Other titles: Mixen met impact. English
Description: First edition. | New York, NY : Routledge, 2019. |
"Previously published in Dutch by Edusonic, 2016." |
Includes bibliographical references and index.
Identifiers: LCCN 2017051305| ISBN 9781138080881 (hardback: alk. paper) |
ISBN 9781138080898 (pbk: alk. paper) | ISBN 9781315113173 (ebook)
Subjects: LCSH: Sound recordings—Production and direction. |
Popular music—Production and direction.
Classification: LCC ML3790.O49 2019 | DDC 781.3/4–dc23
LC record available at https://lccn.loc.gov/2017051305

ISBN: 978-1-138-08088-1 (hbk)
ISBN: 978-1-138-08089-8 (pbk)
ISBN: 978-1-315-11317-3(ebk)

Typeset in Sabon
by Florence Production Ltd, Stoodleigh, Devon, UK

This book is also available in Dutch through Edusonic.
For more information, visit www.edusonic.nl

Contents

Acknowledgments

A book doesn't come into being overnight. The following people and organizations have played an indispensable role in the creation of the stack of paper you're holding right now. Some of them have even shaped me, both personally and professionally. First of all there is Eelco Grimm, who provided me with knowledge and inspiration, and who was the first to offer me a chance to publish my writings. Hay Zeelen taught me to listen, gave me notepads full of feedback on my mixes, and encouraged me to keep improving myself. My collaboration and friendship with Wouter Budé has given me years of artistic challenges, and all the unique musicians I've been lucky enough to work with are the reason why I'm a mixer. The motivation to keep writing was sustained by the Music and Technology classes I taught at the University of the Arts Utrecht, which is where many of the ideas in this book took shape. Jean-Louis Gayet was kind enough to publish my articles (which formed the basis of this book) in *Interface*, the magazine edited by Erk Willemsen at K18 Publishing. Meanwhile, Erica Renckens taught me a lot about syntax and form. Niels Jonker and Mike B at Edusonic Publishing motivated me to finally finish this book so they could publish it in the Netherlands. Gijs van Osch then took on the enormous task of translating *Mixen met Impact* into *Mixing with Impact* (and did a great job), so Lara Zoble could publish it at Focal Press. And last but not least, Tünde van Hoek made me go through the various stages of creating this book with a smile on my face.

Preface

This book is about mixing. About the process I completely lost myself in while making my first demo cassettes. I have gained many insights since then, but every new mix is still almost as big a challenge now as it was then. Mixing is a competition with yourself that only lasts as long as you can sustain your concentration and objectivity. That's what keeps mixing challenging for me: I manage to accomplish more and more before time runs out. I can immerse myself into the music deeper and deeper. What hasn't changed over the years is that mixing can drive me to despair. But the satisfaction of all the puzzle pieces falling into place is still indescribable.

Travel Guide for This Book

Just like you return to the basics every now and then during the mixing process, this book lends itself to occasionally browsing back and forth between chapters. You will notice that everything is connected, and that the greater meaning of some chapters will sometimes only become clear after reading all the other chapters. However, the order in which the different topics are discussed was chosen for a reason.

After clarifying what your challenge is in Chapter 1, you will learn to listen in Chapter 2. You'll then be ready to shape the four basic components of your mix: the balance (Chapter 3), the frequency distribution (Chapter 4), the dynamics (Chapter 5) and the space (Chapter 6). Taking this to the next level involves creating a mix that conveys a strong identity (Chapter 8) and manipulating the sound to reinforce its characteristic qualities (effects, Chapter 9). Just like a good story, a good mix builds suspense (Chapter 10) and feels like a whole (Chapter 12). Creating this is easier if you first shape your own instrument and play it like a musician (Chapter 13). At the end of the day, you will hopefully finish a mix that you want to send out into the world (Chapter 14), after which you start over at Chapter 3 the next day. If you go through this cycle enough times, the technique will become second nature, and the biggest challenges will be mental (Chapter 15) and social (Chapter 16).

Chapter 1

The World of Mixing
About the Profession and This Book

Mixing is not the same as making music. It is the art of translating music from separate components into a coherent whole. These separate components might be recorded on individual tracks, they might enter your mixing desk live, or you can generate them yourself using all kinds of equipment.

These components will hopefully contain the soul of the music, which you try to communicate to the listener. You're telling the story of the music with the sound that's available to you. Mixing is not an autonomous art form. Without musical ideas, arrangements, recording and production there would be nothing to communicate.

1.1 Who Is the Mixer?

As a mixer you have to be crazy enough to want to work long hours, struggling to develop other people's ideas to their fullest potential. No matter how brilliant you think your own input is, if the musicians, producers, record companies and audience you work for disagree, you'll need to give in. Mixing is not a job for people with big egos, but to complete a mix to satisfaction, you do need a lot of self-confidence, social skills and persuasiveness. If you don't dare, if you don't believe in your own abilities, the mix is doomed from the start. And if you don't share your excitement and passion about where you want to take the music and why, people won't give your ideas a chance. To get to the point where you can make something you fully—well, almost fully—believe in, besides a lot of experience you mostly need a lot of enjoyment. If you're not having fun, if you don't enjoy the music you're working on and can't be proud of your contribution to it, you'd better start doing something else. Mixing has a lot of technical and organizational aspects, but in the end the difference between a kind-of-nice mix and a brilliant mix is made by your creative decisions. Feeling what the music needs and not being afraid to make this work any way you can, that's what it's all about. Over time and through experience, you can be sure to always deliver a final product that's technically satisfactory, so you can focus more on the creative side.

1.2 Why Is Mixing Necessary?

As soon as you start using electronics to record and reproduce music, you will need a certain degree of translation to make this reproduction musically convincing. Because electronics never do what it says on the box. You work with imperfect systems and concepts, but if you can handle the limitations, the music doesn't need to suffer. This is why even classical music recorded with two microphones often needs some postproduction. Not so much to change the music, but to prepare it for listening through loudspeakers. This preparation goes further than you might think. Because besides the fact that speakers aren't perfect, the entire situation in which you listen to the music is completely different. The excitement of a live performance, the buzz among the audience, the high sound levels and the spectacle of the stage and lights: all these together create the impact you have to do without at home. Many composers, songwriters and producers are only interested in this impact, so it's understandable that they try to find ways to translate it to your living room or your earbuds. In classical music, this often means that details are magnified and the overall dynamics are reduced. This way, it's easier to hear all the elements of the music under conditions less ideal than in the concert hall. A fragile solo that's 'highlighted' a bit in the mix can grab you more, even if you're not sitting on the edge of your seat to see the lone soloist on stage.

In less conservative genres, this magnifying of musical information can go much further, with the mix becoming a formative part of the production. In some productions, the mix is so determining for the character of the music, that the mixing technology used in the studio is subsequently taken to the stage. Then it needs to be translated again for the studio mix to work in a live setting, because sound that you hear but don't see can ruin your entire live experience. This transition will also require a mixer with talent and a feel for music. In summary, every time the music is brought to a new setting, a new mix is needed, and it depends on the musical context how far mixers can go with their manipulations. Sometimes it's only about playing with the listener's subconscious, other times you attract all the attention with unique sounds and huge mixing contrasts.

1.3 What Is the Challenge?

It seems so easy to put together a jigsaw puzzle of musical elements. But mixing is not the same as assembling a thousand-piece picture of an alpine meadow, because the shape of the puzzle pieces is not fixed. All the pieces that you shape and put into place will affect all the others. This makes the mixing process very hard to understand. If you don't pay attention, you can keep going around in circles because every next step could ruin your foundation. Mixing is an iterative process: you create a starting point, go through a number of steps, and then revise the starting point based on your findings. This book can be read the same way: only after you finish it will you understand some chapters as parts of the larger whole, and you will get what they mean when you read them again some time later. The fact that it is never finished—and therefore, that you never stop learning—is what makes mixing so much fun. Your frustrations in one project are your drive in the next.

Having said this, there are conditions you can set as a mixer to make your life easier (a great deal of this book will deal with those). Many novice engineers tend to confuse creative freedom with having such an easy life. Yes, at first you will encounter little friction if you present yourself as a mixing expert and send the musicians home while you make all the hard decisions yourself. And even though you really do have more mixing experience than the band, this working method is actually not so easy. Because now you have to think about every technical, musical and productional choice yourself. This is a recipe for indistinct mixes that lack a clear focus. I discovered this when I made the transition from doing everything myself in my own studio (recording, producing, mixing and often mastering as well) to working for other producers. Initially, it was annoying to deal with all this negative feedback and to defend every mixing choice, but the mixes were all the better for it.

For you as a mixer, technology is your friend but your enemy as well, because before you know it the importance of technical perfection will start to dominate your actions. That's bad news for the mix, because technology should always be subservient. So someone who has a clear plan and who pushes you to make it work musically can be a welcome counterbalance.

The more I am forced in a certain direction by already recorded sound modifications and strong opinions on what the mix should or should not sound like, the better I perform as a mixer. It can even be difficult to improve the feel of the raw recordings, but that's what makes it challenging. Mixers love limitations: nothing is more annoying in a process that has to come about within a certain time frame (whether this is live or in the studio) than a postponed decision. No, it's not nice to have an extra DI track of everything, or four stacked kick drum samples or three alternate takes to choose from. Do something and live with it, then the mixer can add much more to the project in the limited period of time that's called concentration.

This brings us to the main difficulty in mixing: there is no truth, and your perception of the truth is constantly changing. You might wonder why there are still people who want to mix, because this part of the process is a real mindfuck that can drive you to sleep deprivation and insecurity. The first time you hear a piece of music is priceless. It will never make the same impression on you again: from now on, your perception of the music will constantly change. You will get used to it, you'll get attached to its current form, discover things in it that no one else will ever hear, try things that don't work and make you insecure, you won't recognize obvious problems through sheer habituation, you will adjust your reference of what it could be, and you'll look at the clock and realize it's high time to choose a final approach. And then you think it's finished, but how on earth will it sound to someone who's hearing it for the first time?

This book will help you to slowly get a grip on the many processes that are tangled up during mixing, so in the end you'll be able to utilize the limited window of objectivity you've been given as a mixer as efficiently as possible.

1.4 On Rules

For all the information about mixing that you find in this book (except maybe the technical background), I can give examples of people who do the exact opposite, and with success. Your creative process is a very personal thing that I don't want to change with this book. I do hope to nourish it with concepts that are universally useful, knowledge of your gear and how it relates to music, and stories of methods that happen to work well for me. Together, these are the tools for finding your own solutions through lots of practice. Because you will run into the same problems I run into every day, but I still can't tell you what you should do in those cases. All I can do is dust off the only universal mixing rule there is:

Find a way that works, no matter what.

Who knows, the ways you come up with might one day grow into a *signature sound* that will make people want to work with you for years to come.

Chapter 2

Listening
Everything Is Relative

Anyone with a computer, some software and a pair of speakers has access to a studio that's more than adequate. In fact, some world-class productions were mixed with the exact same tools that are at the average home producer's disposal. So theoretically, you could compete with the very best if you had the chance to work for top artists. It all comes down to the choices you make.

And this is where things get tricky, of course. You will need a combination of talent (which this book unfortunately can't give you), know-how (which this book will hopefully contribute to), and reliable monitoring. The latter is the first topic to be discussed here, because every mixing decision you make is based on what you hear. Of course, this depends largely on the characteristics of your room and your speakers, but how you use them is the biggest factor. That's why we will start with our finest tool, and all its peculiarities: the ear.

2.1 Perception

During mixing, you make decisions about the relationship between individual sounds, the use of the frequency spectrum, and the dynamics. Naturally, you can only rely on what you hear. But what you hear—and how you interpret this—is dependent on more factors than you might expect at first.

The human auditory system is like a set of microphones, hooked up to an impressive signal processor, which interprets the incoming sound. The main difference between a biological and an electronic ear is whether the sound is transmitted linearly or not. A microphone is very capable of linearly converting sounds within the human hearing range into electricity. In other words: the frequency spectrum and dynamics of the electric signal transmitted by a microphone are virtually the same as the frequency spectrum and dynamics of the source sound. Our ears, on the other hand, don't respond to sounds in a linear way: the auditory response is dependent on the properties of the source sound (loudness, dynamics, frequency

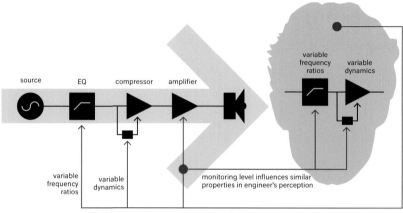

Figure 2.1 Perhaps the hardest thing about working with sound: you manipulate properties of the source sound, while similar properties are simultaneously being influenced in your perception of that sound, depending on the monitoring level you use.

characteristics) and on the duration of the ear's exposure to the source sound. And it just so happens that these are the very properties you manipulate during mixing (see Figure 2.1). So if you want to make reliable mixing decisions, you have to keep the behavior of your ears in mind.

2.2 Your Ear Is a Compressor

Adaptation

Your auditory system is aimed at distilling as much relevant information (such as voices, locations of prey or signs of impending danger) from your surroundings as possible. Sometimes crucial information is quite hidden: it's only barely audible above the background noise. Your ear automatically adjusts itself to these conditions by shifting its own dynamic range to the area where the information is located. This process, which is similar to the way pupils respond to the amount of light that's coming in, is called adaptation. An extreme example of this is that people in a perfectly quiet environment will eventually start to perceive their own heartbeat.

Adaptation works for both faint and loud sounds: a faint sound can be perceived as louder, and a loud sound can appear more faint. So adaptation works just like a compressor, but a lot slower. You'll probably know the feeling of walking into a venue during a loud concert: after five to ten minutes, the sound doesn't seem as loud as when you just came in. And when you bike back home, it also takes a while before you hear the sound of the wind in your ears getting louder again. And that's a good thing, because if the adaptation mechanism reacted quicker, it would constantly intervene in musical proportions, like a compressor. Adaptation can be an obstacle when making critical mixing decisions, but it doesn't have to be if you adjust your working method to it.

During mixing it's customary to vary the monitoring level, so you can evaluate the mix at different sound levels. The best mixes sound balanced and impressive whether they're played softly at home or blasting at a club. But if you vary the level too fast, your ears don't have time to adapt. A common mistake is returning to an average monitoring level after you've been listening at a high level for a while, and deciding that you need more compression. Because the details you heard before now seem to have vanished, it makes sense to bring them up by decreasing the overall dynamics. But if you had held off your judgment just a little longer, your ears would have adapted to the new listening conditions, and all the details would have automatically reappeared.

Another pitfall is turning the sound up for just a moment, which will immediately make you impressed with the impact of your mix. But if you listen a bit longer at this high level, you will start hearing the same problems in your mix. So it's very important to fully adjust to the new monitoring level. A rather nostalgic engineer said in an interview that what he missed most about using analog tape was the rewinding part, because this would create a necessary moment of rest before he could start evaluating the mix afresh. In a modern setting, it's a good idea to create these moments of rest yourself, for example by having short breaks. After changing the monitoring level, it can also be very refreshing to first listen to a production front to back at the new level before you make a single adjustment. This will allow you to form a new frame of reference (to gain perspective on the project as a whole), which can then become your new starting point.

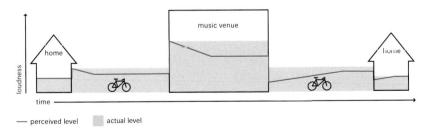

Figure 2.2 The dynamic range of your auditory system—and therefore the perceived loudness (the red line)—adapts to the ambient noise. However, it takes a while to get used to new conditions, which is why the concert seems much louder in the beginning than at the end, and the traffic noise on the way back seems much softer at first than when you're almost home.

Protection

Exposure to high sound pressure levels can damage your ears. That's why we have a built-in protective mechanism that's activated above 80 to 85 dBSPL. This mechanism, which is called the acoustic reflex, causes contraction of the middle ear muscles when the threshold is exceeded. Or when you start talking, because the acoustic reflex also suppresses the sound of your own voice, so you can keep listening to your surroundings while you talk. These muscle contractions hamper the transmission of sound in the ear, causing compression with a ratio of about 3:1. However, this compression mostly affects low frequencies, so it also affects the perceived frequency spectrum.

Unlike the adaptation mechanism, the acoustic reflex reacts much faster (with an attack time of 50 to 100 ms), so it can considerably affect the perception of musical proportions. Therefore, it's not a good idea to determine the amount and type of compression in a mix at a very high monitoring level, because then you'll get the compression inside your ears as a bonus. If you listen to an overly dynamic vocal recording at an average monitoring level, it seems to pop in and out of the mix, but at a loud level it's integrated into the music.

This effect can be very deceptive, and it's one of the reasons why most mixing engineers balance their mixes at a medium or low monitoring level. However, a high monitoring level can be very useful if you want to determine whether you've gone too far with the use of compression and limiting. Combined with the compression of the ear, the low frequencies of an overly compressed mix will be substantially attenuated, and the sound image will appear flat and shrill. The way in which the ear's protection mechanism responds to a sound image is one of the reasons why varying the monitoring level can give you a lot of insights into the characteristics of this sound image.

> A protection mechanism is there for a reason, and it's anything but infallible: the longer it's activated and/or the louder the sounds, the higher the chance of hearing loss. So listen only briefly at a high monitoring level to gather the information you need, and then return to your reference level.

Fatigue

It's better not to make the most critical mixing decisions at the end of a long day. Besides the fact that it's hard to remain focused, there's also a change that occurs in the ear after prolonged and/or heavy exposure to sound. The level of your hearing threshold (the lowest sound intensity you can perceive) gradually shifts upward. As a result, it will be more difficult to hear soft sounds, and details will disappear from the sound image. Because of this, making decisions about balance and dynamics will become increasingly hard with time. This effect is called a temporary threshold shift.

Initially, a temporary threshold shift won't make working with sound impossible, but it can cause you to end up in a vicious cycle that will eventually make it impossible for you to work, and in the worst case it will give you permanent hearing loss. When a threshold shift causes the soft details in your mix to disappear, your will need to increase your monitoring level to hear the same sound image as before the threshold shift. But this will make the loud parts of your mix even louder, and it will put an extra strain on the protection mechanism of your ears, which complicates decision-making in mixing. After that, the threshold shift will only increase due to the higher monitoring level, making the problems gradually worse.

What Can You Do to Prevent This?

First of all, it's important to take short breaks regularly, and to stick to a fixed reference monitoring level (which you can only deviate from for short periods of time). This is the only way to stay focused and reliable if you work with sound eight hours a day. As soon as you notice your ears getting tired (with some experience, you can learn to sense this pretty well) it's time to stop and call it a day. If you can't do this because of an approaching deadline, don't be tempted to increase your monitoring level. Accept that you can no longer hear all the details, and don't try to solve this by using more compression, for instance. Trust the balance you made before your ears got tired. Hold on to your rhythm of working at the reference level, and use lower and higher monitoring levels only to briefly check your decisions. And after an extremely long day of intense sound exposure you should try to give your ears some more time to recover.

There are tables that show the length of time and level of exposure at which there's a risk of permanent hearing loss. Each country has set its own standard, and internationally these standards differ by a couple of dB. It's hard to say exactly at what level sound exposure becomes harmful, and this can be different from person to person. Both the duration and the amount of exposure determine the risk. For example, being exposed to 80 dBA for eight hours is just as damaging as two hours of 86 dBA (see Figure 2.3).

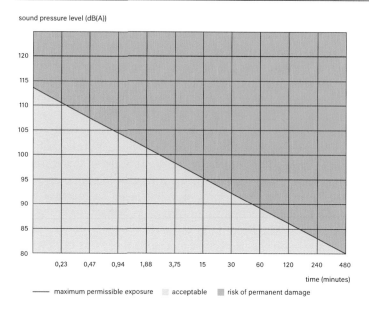

sound pressure level (dB(A))

—— maximum permissible exposure　　▢ acceptable　　▪ risk of permanent damage

Figure 2.3 Permissible noise exposure. Rule of thumb: reducing the duration by half means you can raise the intensity of the exposure by 3 dB, without increasing the risk of hearing loss. Exposure in the area above the red line can cause permanent hearing loss. The 80 dBA point for eight hours of exposure is based on the European noise standards. Other standards have slightly different values: the American standard sets this point at 85 dBA, for instance. But all the standards fall within a pretty narrow range, and it's clear that regularly exceeding the standard by more than a few dB is almost guaranteed to lead to hearing loss.

2.3 Your Ear Is an Equalizer

Frequency Response

If you listen to music and you keep turning the volume down, at the very end only a couple of things are still audible: the sounds in the upper mid-range. At that level, the bass has vanished into thin air. This is because the ratio between the different frequencies you perceive is directly dependent on the sound intensity at which you perceive them. At a very low level, your ear is much less sensitive to low frequencies than to mid frequencies. Based on experiments, so-called equal-loudness contours have been created: graphs that show at what intensity test subjects perceive each frequency as equally loud. These curves are also known as Fletcher–Munson curves, named after the researchers who first conducted this study in 1933, though they have been slightly refined since then (see Figure 2.4).

A Mix Is Won in the Midrange

The ear is most sensitive to midrange frequencies, not coincidentally the area where most of human speech is located. At higher loudness levels, the sensitivity to high and especially low frequencies increases, while it remains the same for mid frequencies. This continues until the point is reached where the acoustic reflex from the previous paragraph starts to temper the relative sensitivity to low frequencies again. And at the highest loudness levels, the hair cells are damaged (temporarily or permanently), which causes the relative sensitivity to high frequencies to decrease. The most important thing to learn from these curves is that the low and high ranges will always sound different, depending on the monitoring level. If your mix works in the midrange—if all the instruments are distinguishable and nicely balanced in this area—there's a good chance your mix will sound acceptable at any monitoring level. That's why many mixing engineers regularly switch to small speakers that only reproduce sounds in the midrange. To judge the right amount of high and low frequencies to complement this midrange balance, it helps to vary the monitoring level of your full-range system. This way, you will notice if the extremes are getting over the top.

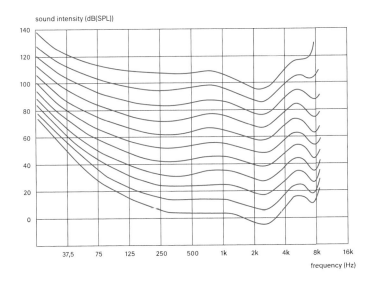

Figure 2.4 Equal-loudness contours: all the frequency points in one line are perceived to be equally loud, even though their acoustic energy can differ up to 70 dB. The graph clearly shows that the highest and especially the lowest frequencies are perceptually louder at a higher sound intensity.

Getting Used to Your Own Bad Mix

If you listen to the same sound long enough, your auditory system will become less sensitive to it. It's interested in perceiving changes, so especially long, sustained bass notes or resonances will automatically attract less attention after a while. For example: you hear the hum of the refrigerator when you enter the kitchen, but after an hour you hardly notice it anymore. This phenomenon is different from adaptation in the sense that it's just the refrigerator that becomes softer in your perception, not the entire sound image. That's why it's called frequency-dependent adaptation. Try to pay attention to it during your next mix: long, sustained sounds may seem loud when they make their entrance, but after a while they become less and less apparent.

You can even adapt to the sound of your own less-than-great mix. The fact that the frequency spectrum is out of balance will automatically become less noticeable if you listen to it long enough. It doesn't help that you know which individual components make up your mix: you can hear all of them, because after a couple of hours, your brain can sift them out from your mushy mix. Such an unbalanced perception of the frequency spectrum can be prevented by taking regular breaks to allow your ears to recover. Besides this, try not to focus on short sections of a production (don't use the loop function), but listen to the project as a whole, so you always hear the onset of individual sounds, instead off 'dropping in' halfway. The onset of a sound determines a great deal of its perceived loudness, and only when you listen to a production in its entirety can you judge the balance of individual elements in their context. Section 15.4 will focus on ways to stay fresh while mixing.

Hearing Accurately

You can distinguish different frequencies thanks to a membrane in your ear that can move along with a gradually decreasing frequency from beginning to end: the basilar membrane. It's shaped in such a way that each frequency causes the greatest movement in a different part of the membrane. The movements of the membrane are picked up by hair cells that are connected to a nerve, which transmits an electrical signal when its hair cell starts to move. This mechanism is supported by the brain, which significantly increases the resolution of the system through analysis of the nerve signals.

The mechanism is most effective at low sound intensity levels, because high-intensity sounds will not only trigger the hair cells corresponding to their own frequency, but the adjacent hair cells as well. As a result, the frequency-resolving power of the ear is diminished at high sound intensity levels. This explains why it's harder to hear if an instrument is in tune at a high monitoring level. So make sure you don't set your monitoring level too high when you judge the tuning of instruments, the tonal precision of vocal recordings, or the amount of pitch correction to apply.

2.4 Tuning In

So far I've discussed the mechanisms your ear automatically activates, outside of your control. In a very different category are the phantom images you create yourself. It happens to me at least once a month that I grab an equalizer knob, start turning it and hear the sound change as expected . . . until I notice that the equalizer is in bypass mode. Because I have an expectation of what's going to happen, I hear it too. This mistake is symbolic of a bigger problem: when you focus on selective aspects of your mix, you will automatically hear them better. It's actually possible to 'tune in' and focus your hearing on a single instrument within a complex whole.

This is what's so hard about mixing: you have to listen to the instrument you're manipulating, but you shouldn't focus on it so much that you start to perceive it out of context. Therefore, I try to listen to the imaginary position of an instrument in the whole, instead of just listening if it's loud enough. I try to place it at a certain distance, and less at a certain volume level. If you force yourself to listen to the relationships within the whole this way, it will help you to prevent your perception from drifting off track.

Objectivity

We can easily fool ourselves if it's in our own interest. If you've just hooked up overly expensive speaker cables, you can bet your life that you'll hear a difference, simply to justify the purchase. The influence of a cable is something you could test scientifically, if you want to have a definite answer. If only it were that easy for the ideas you try out during mixing, because these ideas constantly affect your perception: sometimes you come up with such a beautiful concept that you literally can't hear that it doesn't work. There is one powerful remedy for these mirages: time. Taking some distance and then listening again is usually all you need to make an objective judgment.

2.5 A Fixed Reference

When looking for a good mix balance, you want to work by gut feeling. In order to do so, you should be able to rely on a mental picture of what a good balance sounds like: your reference. This reference should be linked to a specific sound level: it's no use comparing a mental picture that works at a high sound level to a mix that's coming from the speakers at an average level. That's why it's a good idea to set a fixed monitoring level. This is called calibrating your system. Of course you don't need to work at this level all the time (by now you know how useful it is to change your monitoring level every now and then), as long as you can always fall back on it to compare your mix to the reference in your head.

How high should your reference level be? In any case, it should be a level you can work comfortably with all day, and without tiring your ears. You will feel what's the right level for you, and then it's just a matter of putting a mark next to the volume knob or saving a preset. A good starting point many engineers will eventually reach is the fact that pink noise with an average (RMS) level between –20 and –14 dBFS at the listening position produces a sound intensity of about 83 dBC (reproduced by one speaker). Then they will mix at that level as well: the average acoustic loudness of their mix will fluctuate around that level.

The RMS value that you use for calibration depends on the intended dynamics. If you work on pop music with little dynamics and a high RMS level, you calibrate your monitors at this higher RMS level to avoid working at an acoustic level that's tiring your ears because it's too high. But if you work on classical music with a lot of dynamics but a much lower average RMS level, you calibrate the monitors at this low level. This method is similar to the calibration system that's been the norm in

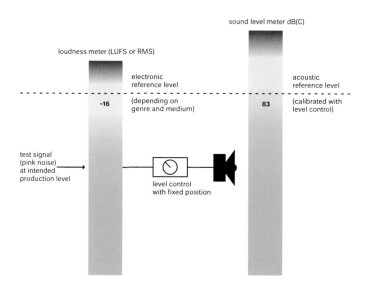

Figure 2.5 If you set your mix at a specific reference level on the loudness meter and match this level with a fixed acoustic level, you will always know what to expect. The loudness level can be measured with an RMS meter, but an integrated LUFS meter that measures the average loudness of your mix as a whole is even better. More information on these meters can be found in section 14.4.

the film industry for years, and it also resembles the K-system for music studios, which was later introduced by Bob Katz.

From the perspective of a mastering engineer (which just so happens to be Bob's profession) a calibrated system is ideal. It provides a solid base for assessing the dynamics and frequency spectrum of a mix, because mastering engineers know exactly what to expect of a certain type of production at their usual monitoring level. For mixing engineers a calibrated system is a useful aid, but it's not the holy grail. Unlike mastering engineers, mixers don't work with material in which the internal balances are already (mostly) fixed. You first have to go through an entire process before there can be a mix with an RMS level on a meter that

---

Decibel

When you work with sound, you will come across decibels all the time, and there's a good reason for this. Because in sound it's all about ratios, and the easiest way to express these is with a unit of ratio like the decibel. For example, when talking about the strength of a sound signal, you can express this in absolute terms: 'This signal causes a peak voltage of 2 volts.' However, this absolute number only means something in relation to the reference of your system. If your system clips at 1 volt, you know you'll need to attenuate the signal by a factor of two to prevent this. This can also be expressed in decibels: reducing the strength of the signal by half corresponds to a decrease of 6 dB. The advantage of the decibel is that it's easy to express a huge dynamic range with it. The difference between the faintest and the strongest sound human hearing can perceive is enormous: the strongest sound causes an increase in air pressure that's one million times greater than the faintest sound. Because the decibel is a logarithmic unit, even this huge ratio can be clearly expressed with it:

$$\text{Ratio (dB)} = 20 * \log\left(\frac{1,000,000}{1}\right) = 120 \text{ dB.}$$

So the strongest sound is 120 dB louder than the faintest sound.

Another advantage of decibels is that they make calculations easy. If you first amplify a sound by 2 dB and later by an additional 5 dB, the total amplification is 7 dB. If you had to calculate this with a non-logarithmic unit, you would have to multiply it twice by large numbers.

actually means something. But when you get to the final stage of the mix, the system will start to mean something again. For example, it can be a good indicator when you notice that your mix is around the reference level on your meter, but you don't feel a lot of acoustic impact. Then you're probably wasting too much energy on sound you don't hear.

2.6 Taking Professional Care of Your Ears

Your senses are built to perceive as much relevant information as possible. They adapt to the circumstances. Therefore, your auditory system has a variable character, plus it's prone to fatigue and overload. If you want to depend on it, you have to give it a chance to do its job reliably:

- Make sure the sound level your ear has to adapt to is not so high that this adaptation permanently distorts your perception.
- Use a fixed reference for your monitoring level that suits the type of production (genre/application) and your activities (recording/mixing/mastering).
- You can change the monitoring level briefly to check specific aspects, but make sure you give your ears enough time to adjust to the new level.
- Take regular breaks, and then listen to a production front to back at the reference level to hear balance mistakes. Always create the final balance at the reference monitoring level, preferably with fresh ears at the beginning of a new day, or after a break.
- Avoid overloading your ears by limiting your working days to eight hours as much as possible, and stick to the reference level with discipline, even when you're tired.

Besides this, any other way you can come up with to keep some perspective on a project is a bonus. I once read in an interview with engineer Chuck Ainlay that he always has a small TV on the Sports Channel during a mixing session. This would take his mind off the project every now and then, allowing him to keep a fresh perspective on it.

Chapter 3

Laying the Foundation
Creating a Starting Position

There are no rules for mixing a project. Some mixers turn on all the channels at once and start moving faders, while others begin with the drums on solo and gradually add the rest of the elements. The first approach immediately gives you a sense of context. It makes it easier to hear where the sound of a piece of music could go. However, it can make it more difficult to analyze problems in the mix, because everything is loaded onto your plate at once. Conversely, the second approach allows you to focus on details, with the risk of losing sight of the big picture.

Which approach works best in your situation depends on a single question. How much do you already know about the music you're about to mix? If you have recorded it yourself and you're the producer, there will probably already be a strong sense of direction for the mix, and ideas of how to achieve it. In this case, a detail-oriented mixing approach can be a fast and effective way to create the sound you have in your head. But if you've never heard the music before, this approach is pointless. If you don't know what role the bass drum will play in the mix, why would you put it on solo and start working on it?

3.1 Before You Start

Step 1: Context

Before you can start mixing, you have to figure out how the arrangement works. Once you understand the role of each element in the whole, it's easier to come up with a suitable treatment for a particular element while you're mixing. Therefore, it's a good idea to first listen carefully to the rough mix made by the person who recorded and/or produced the project (even if that's you yourself!). It won't sound perfect yet, but in most cases it will immediately show you how the arrangement was intended. Which elements have a leading role and which ones a supporting role? Which elements provide the rhythm and which ones fill the space?

Which elements interact with each other? What gives the music energy and identity? What makes the composition special, what gives you goosebumps? Is the atmosphere grand and compelling or small and intimate? Should it make you dance, or is it more 'listening music'? All of these indications will help you make mixing decisions with the big picture in mind.

But what if there's no real rough mix to speak of, or the rough mix completely fails to convey the arrangement's potential? Then you must set out to learn the content and potential impact of the music yourself: you'll have to make your own rough mix. You do this by quickly finding the rough proportions that work well for you, or not at all. It can help to impose some restrictions on yourself during this process. Cherish the moment when you still have a fresh perspective on the project as a whole. The time when it's still exciting to hear a new piece of music might be the most precious moment in the entire mixing process. The ideas you gather during this early stage are often the most creative and decisive for the direction of the mix. So when you're creating the rough mix, the trick is to not endlessly tinker with details, but to quickly set up a rough balance that conveys the essence of the music and that feels good to you. Ideally, the second time you play the music your faders should already be in an okay position, and you should have an idea for a suitable approach.

Step 2: Organization

In the past, twenty-four tracks for a production seemed like an unprecedented luxury, but today it's peanuts. Eighty tracks in one project are no longer an exception, and a mixing engineer will have to wade through all of them. Therefore, it's vital to organize the mix in categories. Nobody can store eighty music sources in the back of their head and intuitively grab the right fader when there's a problem. For this reason alone, it's a good idea to balance similar tracks and group them together in a bus. Sixteen smartly chosen stereo buses are enough to control almost all the proportions of an eighty-track mix. Sixteen is a number that's a bit easier to remember, that can be clearly displayed on a single screen, and that's easy to control with an average mixing console or controller. Ironically, this is the exact same thing the engineers of the analog era used to do when they reached the end of the available tracks: material that would easily take up eighty tracks today was partly mixed down and then recorded on sixteen or twenty-four tracks.

How do you categorize the buses? It would seem obvious to create instrument groups like drums, strings and synths. But you can also base the categories on other criteria. For example, you can send all the instruments that make up the low range of a mix to one bus, or all the instruments that together create the groove. You can group all continuously sounding (legato) instruments or short (staccato) sounds, downbeat or offbeat parts, parts that belong in the foreground or the background, and so forth. This way you can control the entire character of the mix with just a few faders. With a single movement you can determine the amount of low end, rhythm or filling. Another advantage of using buses is that you can manipulate several parts at once by applying compression, equalization (EQ) or reverb to the entire bus. Especially when parts need to have a certain coherence, this can be very interesting. For instance, when a vocal part has to merge with a hectic guitar part. Compression in particular can really mold these parts into a single whole.

Mixing is a continuous search for solutions to all kinds of problems. It's like solving a big puzzle. If you organize your tools well and keep them within reach, you can try out any idea in an instant. If it turns out that it doesn't work, you will know right away and you can move on to the next idea. This might sound like stating the obvious, but even if you can immediately activate each tool in the form of a plugin, there is something to be gained from organization. Take reverb and delay for example: if you start your mix by creating a standard setup of a number of effect sends that all result in a different depth placement, you can quickly find out which depth perspective is the best fit for a certain part. This also goes for any hardware you might be using for your mix. It's a flow killer if you still have to connect every single device you want to use when you're already in the middle of the creative process of your mix. So make sure everything is ready to go before you start mixing, because this way you'll be more inclined to experiment, which in turn stimulates creativity.

Step 3: Gain Structure

The relationship between the signal levels in the various components of a mixing system is called the gain structure. In a (partly) analog setup, a mixing system often consists of several components, which of course should be well matched in terms of dynamic range. A less-than-optimal gain structure will impair the signal-to-noise ratio and can cause distortion. In a digital mixing system like a digital audio workstation (DAW), thinking about gain structure doesn't seem as much of a necessity. Most DAWs use 32-bit floating point mix engines, which make noise a non-issue. Floating point means that the ceiling and floor are automatically shifted based on the signal that has to be processed. If it's weak, the dynamic range of the mixer shifts downward; if it's loud, it shifts upward. This way, nothing will end up below the noise floor, and clipping of mixing channels is virtually impossible as well. As long as you keep your final product (and your DA converter) out of the red, there is usually nothing to worry about, even if the source channels do exceed 0 dBFS. However, there are good reasons to not let these source channels reach that level, since some plugins do have a fixed ceiling: they clip at 0 dBFS. Especially the plugins that emulate the distortion character of analog equipment only work well if your source signal is at the right level. As with analog equipment, the trick is to find the sweet spot where you add exactly the right amount of coloration to the signal. If you use these levels in your DAW mixer and calibrate your converters, it will immediately become much easier to integrate analog peripherals or a mixing console in your mix. Section 13.3 will elaborate on this subject.

How do you set up good gain structure? A starting point for in-the-box mixes is that you should be able to leave the master fader at zero without the mix bus clipping. To achieve this, it helps if not all the source material peaks at 0 dBFS to begin with, because this will make it harder to use processing such as EQ and compression without overloading the channels. Also, it will force the channel faders too far down during mixing, into the area where they have low resolution, so the slightest movements cause huge differences in level. This makes the automation of faders very difficult, and it's simply impractical. It's important that you don't record at an overly high signal level. If you record at 24 bits, it's not a bad idea to have percussive sounds peak at −6 dBFS, and less percussive sounds can easily peak as low as −12 dBFS. If a recording peaks too high or too low to use your faders in a convenient range, you can adjust the level with the trim control at the top of the mixing channel.

Keep in mind that in mixing, many people will first think of adding, emphasizing and amplifying things. Intuitively, you want to turn things up. So especially when you're setting your initial balance, it can help to deliberately work in a downward direction to accomplish good gain structure. Start with the channel that probably produces the highest peak levels in the mix (usually the drums), and set it at a level that peaks at −6 dBFS on the stereo bus. Then add the rest of the elements to the mix, but resist the temptation to turn up the drums. If something is getting in the way of the drums, you should only allow yourself to fix this by turning other things down. This way, you'll have an initial balance that leaves you some space to emphasize things later on, without immediately pushing the stereo bus into distortion and causing your DA converter to clip. On an analog mixing desk this limit is not as rigid, but the essence remains the same: you want to leave some room to maneuver. And in live mixing, this approach will also help you to avoid exposing the audience to higher and higher sound levels.

Background Information: Gain Structure

Sound comes in three forms:

1. acoustic, as sound pressure;
2. electric, as a voltage or current;
3. digital, as a sample.

In all three forms, the sound level in decibels means something else: a different reference is used, which is usually specified following the dB unit. These are the most common ones:

1. Acoustic: dBSPL. 0 dBSPL (SPL stands for sound pressure level) is the faintest sound the human ear can perceive at 1 kHz. This sound causes an increase in pressure of 20 micropascals. Therefore, the value of a sound pressure measurement is expressed in decibels above 20 micropascals. Sometimes filters are applied to the measurement to adjust the frequency response to that of the ear; this is called weighting. The most common varieties are A-weighting for faint sounds and C-weighting for loud sounds: dBA and dBC.
2. Electric: dBu. dBu is a measurement unit for the nominal (RMS) voltage above or below 0 dBu, which is equal to 0.775 volts. So the value of a signal in decibels can be expressed positively and negatively compared to this reference. For example, the +4 dBu indication on professional audio equipment means that it's expecting a nominal voltage of 4 dB above 0.775 volts = 1.23 volts. Usually, equipment has enough headroom to handle peaks that exceed the nominal voltage. Some synthesizers, vintage effects and consumer electronics use lower nominal voltages, so they don't immediately connect well with +4 dBu-specified equipment. And keep in mind that some equipment is specified in dBV, which is not the same as dBu! 0 dBV = 1 volt (RMS).
3. Digital: dBFS. In this case, the reference is 0 dB full scale: the highest possible sample value. Consequently, the level of a certain sample is expressed in a negative ratio (in decibels) compared to this level.

How can this information help you to make your studio sound better? If the acoustic, analog and digital parts of your studio are well matched, the highest acoustic level will remain just below the analog and digital maximum level. This results in an optimal dynamic range, minimizing the amount of information that ends up below the noise floor of the system. For example, if you know that your microphone preamp is just below its maximum level at +20 dBu (which is often where it sounds best), you can adjust the input of your AD converter so this voltage corresponds to −6 dBFS (a safe margin) at its output.

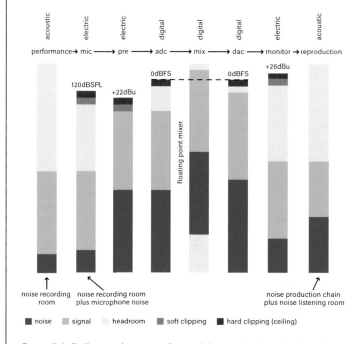

Figure 3.1 Different domains of sound (acoustic, electric, digital) have different upper limits. Some are hard limits (red) and others have a gradual inhibitory effect on the sound (orange). There are lower limits as well, defined by noise. The trick of a good gain structure is to keep the signal (green) within the limits of the domain. It shouldn't distort at the top or fall below the noise floor at the bottom. The room in which you listen to the recording has its own noise, and in most cases this noise is higher than that of the entire production chain, unless you set an extremely high listening level.

3.2 Foundation

The rough mix is ready and you have a pretty good idea of where you want to go with it. When the tambourine starts to get annoying, you immediately reach for fader 12. In other words: time to get started. But where to start? Usually it's not a bad idea to apply a certain hierarchy. If the vocals are the most important element in the mix, it makes sense to first optimize how they come across, and then build the rest of the mix around them. Still, this doesn't always work, because without context it's hard to decide if something is coming across optimally. This is why most mixers first create a starting position in the form of a rhythm section (drums and bass) that works well with the main harmonic instrument.

The rhythm, bass and harmony give meaning to the most important element: the melody. The vocals can only sound loud and in your face, drowning in the space, bright or dull if the other elements provide the perspective. This approach is similar to a painter who first draws up the broad outlines in terms of distance and size before filling in the details. This sketching method is very effective, but it requires quite some practice, because the sound of the rhythm section changes dramatically once you add in the rest of the arrangement. So you should already keep this in mind when laying the foundation.

But how do you anticipate something if you don't know how it will affect the rest? It's important to make the drums and bass pretty solid initially. You do this by exaggerating their musical function a bit. With drums, for example, the power of the rhythm usually has more musical importance than the sound of the room in which the drums are played. You can exaggerate this rhythm by using the microphones with the most attack a bit more, turning off unnecessary channels or using some compression if it doesn't feel stable yet.

Because the rhythm sounds are usually the lowest and highest elements in the frequency spectrum of the mix, it's relatively easy for them to stick out above the rest of the instruments (which mainly occupy the midrange). Therefore, it also helps to make sure there's enough low end and high end in the drums to begin with. At this stage, the drum sound doesn't have to be nice in itself; initially you only pay attention to the musical function. Similarly, the bass should mainly have a supporting role; if it threatens to get in the way of the other instruments too much, you can start by removing some midrange frequencies with an equalizer. These kinds of modifications aren't definitive at all, but if you postpone them, you'll never finish your first draft. By making important aspects stronger and less important aspects weaker this way, you create space in the sound image, which you will need for the rest of the arrangement. The trick is to know how far you should go when you're exaggerating musical functions, because everything affects everything else. If you go too far, the mix will sound fragmented when you add the other elements. If you don't

predominant foundation   weak foundation   supporting foundation

Figure 3.2 The way you position the base of your mix (the instruments that support the music and propel it forward, usually the drums and bass), determines the perspective in which you can place the other elements. A predominant foundation leaves little room for the rest, while a weak foundation doesn't provide enough support and makes your entire mix seem small.

go far enough, you can't even add all the elements before the entire mix starts to clog up.

The rhythmic elements form the imaginary frame of the 'sound picture' you're creating. They determine how much space you have available to position the other elements. During mixing you don't want to change the size of the frame too much, as this messes with the proportions of the entire composition. So it's important to set a fixed starting point, which can serve as the base for the rest of your choices. You can always come back to the starting point later if you want to make small corrections, which in turn will require other corrections in the rest of the mix. The mixing process is often a back-and-forth movement of increasingly small corrections to individual elements, until you finally find the best possible combination. The main thing is to not be too shy when you start laying a stable base and creating space. Otherwise you'll never reach the point where you can hear the entire arrangement at once without things fighting for the available space. For example, overaccentuating the drums a bit and later taking a step back when you hear how they blend with the rest is better than being too careful and never really seeing the big picture because the drums can't support the other elements (see Figure 3.2).

Expanding

Now it's time to add the extra elements to your foundation. It might help to start with the elements that are most important to the music, and work your way back from there. Try to keep thinking back and forth: when you add an element, sometimes the existing base is the problem, but other times it's the element itself. Be increasingly critical of the things you add. If all the key elements are already in the mix and work well together, you want to avoid making huge changes to this balance just to bring out an extra supporting part. As a mixer, you have to choose which elements will play a major role, and which ones a minor one. This includes not necessarily using all the microphones an element was recorded with in your mix, or not using double-tracked parts just because they're there. The more you trim the things that aren't essential, the more space there is left for the important elements to make a big impression. There's a reason why 'less is more' is such a commonly used mantra; you might almost frame it and hang it on the wall above your mixing desk.

3.3 Balance

Of all the concepts in mixing, balance is the easiest one to understand. You just put the faders in the position that matches how loud you want to hear the different sounds in your mix, and you're done. Of course that's just how it works, but knowing how loud you want to hear the different sounds in relation to each other is easier said than done. And after that, it can be pretty difficult to hold on to your planned proportions while you're mixing.

Making Choices

It's often hard enough to make all the elements come out in the big picture, let alone manipulating their proportions a lot without losing all kinds of details. Consequently, you might think there's only one correct position for the faders: the one that results in an extremely balanced mix, in which all the elements have their own place and all the details come out. However, the problem with this method is that it also makes all the elements more or less equal, and there is very little music that actually works that way. Technically, your mix is perfectly balanced, but musically and emotionally it just doesn't cut it.

Virtually all arrangements consist of leading and supporting roles, which should be similarly expressed in the mix. This is the only way for the mix to convey the right emotion to the listener. So the crux in making a good balance is knowing which details are unnecessary, or at least which don't need their own identity in the mix. Of course, this is not an excuse for making an undefined mess of sound with the vocals blasting over it. It's much more subtle: in a mix that's already very well defined, a balance difference of one or two decibels can make a world of difference to the feel of the entire musical piece.

To arrive at this point, you first have to create a balance in which everything has a place. After that, it's a matter of 'killing your darlings': you sacrifice some of your hard-earned details so you can increase the role of the elements that are most important for the musical feel of your mix. This doesn't mean turning instruments off completely; sometimes it's just about minimal balance differences of one or two decibels. This is the most beautiful part of mixing, because with simple changes you can

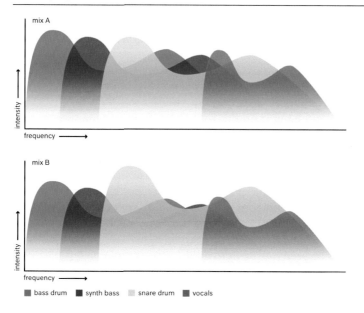

bass drum   synth bass   snare drum   vocals

Figure 3.3 The colored elements represent the frequency ranges different instruments occupy in a mix. In mix A, all elements are well defined, but the parts they play in the music as a whole are completely equal. In mix B, all elements are still audible, but the red and green instruments have to make way for the yellow instrument, which clearly gets the leading role. As a result, mix B will have a more pronounced identity than the playing-it-safe mix A.

make something feel intimate or grand, mild or aggressive, danceable or melancholic, and so on.

At this stage, you will also find out that not everything is compatible. In a heavy metal production, turning up the vocals will make the guitars seem weak. Loud keyboard parts in a hip-hop production make the bass seem small. So a balance is always a compromise, as any orchestra conductor will tell you. Sometimes a bombastic brass line gets priority over the identity of the harp. These are hard choices you can only make if you understand how the music works as a whole. So when you're mixing, it's not a bad idea to take some distance every now and then to contemplate this.

Short and Long

Having a clear vision of the balance you want to make is one thing, but of course you still have to succeed in making this vision a reality. And that's not easy at all, because if you don't pay attention, your ears will constantly fool you. Creating a balance is about continuously assessing sounds based on their loudness. But the outcome of this assessment depends on the type of sound, on other sounds that you hear at the same time, but also on the monitors you use and their level. As you have read in section 2.1, the frequency response of your auditory system depends on the intensity of the sound you are listening to.

But that's not all. For instance, think of a finger snap peaking at –10 dBFS. You'll perceive this as a lot less loud than a trumpet peaking at the same level. This is due to the length of the sound: you perceive long, sustained sounds as louder than short sounds. On top of this, the extent to which you perceive the sustained sounds as louder depends on your monitoring level. If your monitors are very loud, you're more inclined to boost the drums (too much), and if you listen at a low volume, you'll emphasize the vocals and bass instead. So it's very important to check your balance at different volumes in order to find a good average.

And it gets even more complex: sounds that last very long—like a string section, an organ, or a synth pad that's playing long notes— will lose their appeal at some point. In a way, your ear gets used to these sounds being there, and you will gradually perceive them as weaker. Our brain is more interested in things that change than in things that are always there (see section 2.3). This mechanism is a huge pitfall when you're balancing continuous sounds that lack a clear head or tail.

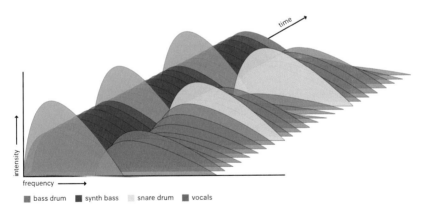

Figure 3.4 Long, sustained sounds (like bass and vocals) tend to fill all the gaps between short sounds (like drums). So, comparatively, the sustained sounds take up a lot of space, and they can easily dominate the sound of the mix if you make them too loud. With the balance in the graph above, you perceive all the sounds as equally loud. To achieve this, you have to balance the low-frequency sounds with a higher intensity than the mid-frequency sounds, and make the short sounds more intense than the longer sounds.

Spectrum Analyzers

When you're mixing, visual feedback can come in handy. Correlation meters, phase scopes, level meters and spectrum analyzers can all provide you with some insight into the technical details of your mix, and tell you if you need to take these into account when you're creating your balance. But be sure to take this information with a grain of salt: seeing is not the same as hearing! Spectrum analyzers in particular can fool you, because they don't tell you enough about the timing of the frequencies they display. And as you can see in Figure 3.4, timing is actually essential to your perception of a balance. A short peak looks very meaningful on a lot of analyzers; it will easily make the spectrum seem full, even though it doesn't sound like that at all. This is the reason why equalizers that let you 'apply' the spectrum of one mix to another don't work. In Chapter 4 you will find even more reasons for this.

You might find yourself listening to the final chorus, wondering: 'Where did the organ go?' But if you turn it up, chances are that it's suddenly much too loud when it kicks in three minutes earlier.

Therefore, a good strategy for these types of sound is to be a bit conservative in the beginning about how loud you put them in the mix, and only highlight them at the moments when they're important to the music (through volume automation). For instance, if you clearly introduce a part when it kicks in for the first time, you can turn it down later. As a listener, you identify the part when it starts and give it a place in your head, so it's easier to keep track of it, even if it's turned down later. If the part is fading into the background like the organ in the third chorus, you can also give it a short boost on the first beat of the chorus. This prevents you from slowly and unnoticeably clogging up your mix.

Mini-Speaker Check

The key to a mix that translates the music well to all kinds of speaker systems is a balanced midrange, because almost all speakers are capable of reproducing this area reasonably well. With some mixes this happens automatically; when you switch from your full-range monitors to a couple of band-limited computer speakers, you can hear the sound becoming smaller, while the musical proportions remain largely the same. However, certain sounds can really throw a monkey wrench into your balance. On large speakers a tambourine can appear nicely proportioned to a warm-sounding bass, but on small speakers, without the low end, it can suddenly attract way too much attention. This doesn't necessarily mean that if you then turn the tambourine down, it will appear too weak on the big speakers. Switching between different speakers is primarily a tool for identifying problems that only become apparent on certain speaker types. To avoid 'translation errors' like this, many mixers choose to do the final balance and volume automation of vocals on small monitors that only reproduce the midrange. Generally, it's easier to translate balances created on small monitors to large monitors than the other way round. Of course, this method is only reliable if you first lay a solid foundation for your mix on large monitors.

Contrast

One reason why listeners perceive recorded music as different from live music is the fact that the available dynamics are greatly reduced. Of course you can capture the full dynamics of a symphony orchestra in a 24-bit recording, but if you want to reproduce these dynamics, you will have to play the music at orchestra level in a quiet room. So it's not surprising that even the purest acoustic recordings are dynamically reduced a bit for the sake of listening pleasure at home. However, this means that an important aspect of the music will lose some of its power. When it comes to classical recordings this is a pity, but there's not a lot you can do about it without affecting the natural character of the recording. But with musical styles that are less attached to the best possible representation of a live event, nothing stops you from compensating the loss of dynamic contrast with another contrast tool. And the mixing balance happens to be the perfect contrast tool!

Up till now, I've talked about the ultimate balance that makes a piece of music sound the best it can. But if music changes over time, why can't your mixing balance do the same? If you succeed at making your mixing balance change along with the dynamic contrasts in the music, you can create the illusion that the dynamics are much greater than they really are. Especially if you use automation to cause sudden changes in the mix balance at key moments, you can give these moments a lot more impact. Of course it doesn't work if you simply turn up the entire mix when the chorus starts; then you're only increasing the absolute dynamics. It's the change in sound that makes the contrast, not the absolute volume change. For instance, a dark-sounding verse with a lot of emphasis on the bass and vocals can segue into a bright-sounding chorus with more emphasis on the guitar, percussion and vocals. A couple of well-chosen fader moves of one or two decibels can have a big impact here as well. You can read more about automation in Chapter 10.

If it Doesn't Work Right Away

In mixing, a good fader balance is more than half of the work. But there are cases in which you can't find a balance that works right away. For example, when an element sounds nice in itself, but it gets in the way of the sound as a whole too much. Or when sounds don't want to 'bind' together, or when they weren't very conveniently chosen or recorded to begin with. You can tackle these problems with the techniques that you'll find later in this book: EQ in Chapter 4, compression in Chapter 5.

3.4 Panning

You wouldn't think so because of their simplicity, but along with faders, pan pots are the most important tools for shaping your mix. A good mix tells a story: not literally like a book, but there are different roles to play, and events for the characters to get caught up in. The listener derives a sound's role in the story from the location of that sound. If it's relatively loud and coming from a clear direction, it will attract attention and almost automatically play a leading role. Faint sounds that come from the same location as the leading role have a supporting role, and faint sounds that fill the spaces between the leading and supporting roles serve as extras.

How does Steeo Work?

You can accurately determine the direction of a sound because you have two ears. For example, sound coming from the right will end up slightly louder and earlier in your right ear than in your left, because your head functions as a barrier in between. Your brain almost flawlessly connects a position to the difference in loudness and arrival time between your two ears. And by slightly moving your head in relation to the source (or when the source itself is moving), you can also distinguish between front, back, top and bottom (partly because your auricle colors sounds from different directions differently). Of course, in a stereo mix the sound always comes from the front, but by controlling the intensity and arrival time of various sounds, you can create the illusion of a three-dimensional image. That's what stereo is: a kind of 'shoebox theater' for sound.

The easiest way to position a sound in the stereo image is to send it louder to one speaker than the other. This is called intensity stereo, and the pan pots that you'll find on any mixing console work according to this principle. It's a pretty convincing way to simulate direction, except for the lowest frequencies. These will reach both your ears about equally loudly anyway, because your head is too small to block them. Also noticeable is the fact that the same element won't sound the same from every direction. Just try to move a sound that was recorded in mono

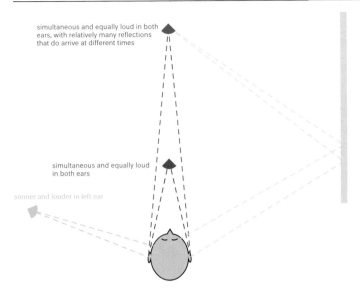

Figure 3.5 Because the same sound arrives at each ear at a different time and with a different loudness, you can hear exactly where the colored sound sources are located.

from left to right. You'll notice that it sounds 'purer' at the opposite ends of the stereo image. That's because you are only using one speaker to reproduce the sound.

In any position between the two extremes of the stereo image, both speakers work to reproduce the sound, and because the sound from two speakers never reaches your ears at exactly the same time, it will seem a bit murky (especially in the highest frequencies). It's not the end of the world, but you do have to get used to this principle when you first start mixing. On headphones, this problem doesn't occur, which is why elements positioned in the center will sound a bit closer and louder than through speakers.

The Third Dimension

Pan pots only influence the proportion of a sound in one speaker compared to the other, but they don't make a sound coming from the right reach your right ear sooner as well. Of course, sound from the right speaker will always reach your right ear a bit earlier than the left. But this time difference implies that the sound always comes from the exact location of the speaker itself. What you really want is that it can also sound as if it's coming from inside or outside the speaker, or from the depth behind the speaker. To achieve this, you will need time differences between the two speakers: you always send the sound to both speakers, but delay it in the speaker opposite to the direction where the sound is coming from.

The longer the delay, the further the sound moves to the side. This principle is called time-of-arrival stereo. Nearly all classical recordings work with time-of-arrival stereo: they use two microphones positioned at a distance from each other to capture an entire orchestra. Instruments in the middle of the front row will arrive at the two microphones simultaneously and with equal loudness. Instruments that are also in the middle but positioned at the back row will be picked up with more reverberation. Because reverb comes from all sides, the effect of it reaching the microphones at the same time is greatly reduced. As a result, these instruments cause more time differences between the two microphones, but because of the random, diffuse character of reverb, you will still perceive them in the middle (simply further back). Instruments on the left will end up in the left microphone earlier and slightly louder than in the right, so you perceive them on the left side. This way, you can record a three-dimensional image with two microphones (like in Figure 3.5, but with microphones instead of ears). When you're mixing material that wasn't recorded this way, you can still simulate this principle with delay. You'll read more on this in sections 6.6 and 9.6.

Action Plan

Where do you position each sound? For classical music recorded with only a couple of microphones this placement is already set, and a good recording won't require a lot of manipulation, but a production with a lot of close miking or synthetic sources is very different. Of course you can move things around randomly until it sounds good, but if you put some thought into it you can make your life a lot easier. First of all, it's a good idea to make a distinction between the parts that play an important role and the parts meant for support or coloring. Luckily, there are no panning police who will come and get you if you throw this advice to the wind, but in general it's not a bad idea to place the important parts in the center of the mix. Someone standing in the front and center of the stage will easily get a lot of attention; aurally it works the same way.

Besides the importance of a sound for the mix, there's another thing to consider: headroom. If you want to reproduce sounds with a lot of low-frequency energy (like drums and bass) loudly, the most effective way to do this is with two speakers. Think about it: if two speakers reproduce the exact same sound (a bass guitar recorded in mono, for example) just as loudly, the two reproductions will add up in the middle between the speakers. This combination will sound louder than what you can achieve with a single speaker at the same signal level. So if you choose to move sounds with lots of low frequencies to one side of the mix, you'll have to turn the entire mix down a bit to keep it balanced. This is why in electronic dance music—where a loud bass drum and bass are sacred—these elements are always placed in the middle.

Once you've decided which key elements will occupy the center of the mix, it's time to fill in the stereo image. A good way to do this is to determine the role of an element in the story. Every time, you should ask yourself the question: 'Should this element be individually audible, or should it blend with another element instead?' For example, if you're mixing a vocal duet, it's nice if you can distinguish two leading roles, so it helps to pan the voices apart a bit to give them each an identifiable place in the story. On the other hand, if you're mixing an extra voice that's only meant to add some harmonic color to the lead voice, it's probably a good idea to pan the extra voice in the same position as the lead voice. This way, you prevent it from claiming a role in the story and making the leading role seem less important.

During this process, you might run into the limits of your mix, when there's simply no room left to give things their own place. When this happens, it can help to take a critical look at the source material, in particular at the sources that were recorded in stereo or sampled, and whether this is really necessary. For example, in a ballad it might work very well to record an acoustic guitar in stereo and pan it wide, but this approach usually doesn't work so well if the acoustic guitar is part of a rock production with three additional electric guitar parts. In this case, a mono recording is much easier to position in the mix. So it's not a bad idea to separate the sounds that can be large and wide from the sounds that shouldn't take up too much space. The latter category can then be pared down, for example by only using one half of a stereo recording in the mix.

Another approach that could work is the exact opposite: sounds that are already wide can be made even wider artificially, so they almost seem to come from the area beyond the width of the speakers. This also leaves more space in the center of the stereo image, which is sometimes exactly what you need. In section 11.4 you can read more about this technique.

Double Check

When you're positioning sounds, it's easy to get used to where they are and lose sight of exactly how loud they are in your balance. Switching your monitors to mono every now and then can help to evaluate your balance regardless of the panning. It works best to fold the mix down to mono and then listen to it on a single speaker. But even a simple trick like temporarily swapping the left and right channels of your entire mix can already help to give you a fresh perspective.

Panning Stereo Sources

An instrument that's recorded in stereo can be harder to pan than a single-microphone recording. If the stereo recording only uses intensity stereo (if the microphones were positioned in the same place, for example in an X–Y configuration), there's nothing to worry about. But as soon as you start mixing material that was recorded with two microphones positioned at a distance from each other, usually the only thing you can do is pan these mics apart completely. If you don't do this, the time-of-arrival stereo won't work anymore, because it requires an undelayed sound coming from one speaker and a delayed copy from the other. For example, if you pan the two halves 30 percent apart, there will be a delayed version of the same sound coming from the same direction as the original (see Figure 3.6).

Besides the fact that this breaks the illusion of stereo placement, it also causes an unpleasant interaction between the original and the delayed copy (you can read more about this in Chapter 7). As a result, the sound can appear thin and undefined, so keep this in mind! It's better to turn one of the two stereo halves down a bit. Or even—if it doesn't destabilize the tonal balance too much—to use only one half as a mono source and pan it where you want it.

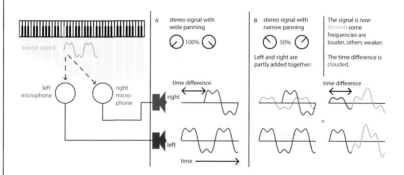

Figure 3.6 Because the piano is recorded using time-of-arrival stereo, it's best to pan the result as wide as possible (A), otherwise you run the risk of changing the sound of the piano and clouding its position in the stereo image (B).

Masking

The difficulty of trying to capture music and reproduce it with two speakers is the fact that there's much less space available than in real life. Just compare the experience of hearing an orchestra play in a concert hall to listening to the same concert at home, recorded with two microphones at your original position in the hall. There's a big chance that you could pick out a lot more detail live than in the recording. At a venue, you use both ears to listen to a three-dimensional sound field, and it's easier to distinguish the sound of the instruments from the reverb: your brain interprets the reverb as less important, and actually makes it softer in your perception (look up the cocktail party effect if you're interested in how this works).

In a stereo mix, the reverb is only in front of you instead of all around you, just like the instruments. Because of this, it's harder for your ears to distinguish the reverb from the instruments. This phenomenon is called masking. It means that perceiving a sound (in this case the instruments) becomes harder or even impossible as soon as a different sound (in this case the reverb) is played at the same time. In a busy mix, masking is usually the biggest problem; it's difficult to make all the details audible at the same time. Masking is a recurring theme in mixing, and I will elaborate on it in Chapter 4.

Panning can be an alternative to EQ or compression as a way to solve masking issues. When two sounds are coming from a different direction, it's much harder to mask one with the other. Therefore, even in mixes that try to stay as true to the original performance as possible, more extreme panning is used than you might expect. This is the perfect way to create a mix with the same amount of detail as you could hear in the audience. In pop music, the approach is much more radical: every inch of extra space you can get is pure profit, because the philosophy behind a lot of pop productions is 'more = more.' So if you only use extreme panning (hard left, center or hard right), you can put more sounds in your mix without having to use an extreme amount of filtering, for example.

On headphones, it's easier to perceive individual sounds in a mix than on speakers, because there is no crosstalk. In this case, crosstalk means that the sound from the left speaker also reaches the right ear, and vice versa. It makes your mix seem slightly narrower on speakers, and it's the reason why you can't reliably judge a stereo image on headphones. Oddly enough, the center of the stereo image is also louder on headphones than on speakers (because even the high frequencies in the left and right channels add up perfectly in phase on headphones). It sounds as if both the extremes and the center of the stereo image are emphasized compared to what happens in between. This is why mixes that translate best sound a bit 'stretched out' on headphones and natural on speakers. If these mixes had sounded natural on headphones, you would lose a lot of detail on speakers.

3.5 Mute Is Your Friend

This is the shortest but perhaps the most important part of this book. It contains only one line, which is also the title: 'Mute is your friend.' Optionally, you can add 'solo is your enemy' to this. What makes this easy-as-pie rule such an important general principle?

The solo button tells you a whole lot about one single element, but nothing about the whole. On the other hand, the mute button tells you everything about the influence of an element on the whole.

When you temporarily remove an element from the context of your mix, you can immediately hear what its influence on the whole was. This is very useful information when you're choosing the best mix approach for this element. But if you put the same element on solo instead, you can't hear its influence on the whole at all. This makes it much harder to determine how to manipulate the element to make it work in the mix.

Solo is mainly useful if you need to find a specific solution for a problem you have already detected, such as clicks or cracks in a recording, an edit, and so on. But during the actual mixing process I try to leave the solo button alone. If you really can't help yourself, try to put the most important elements in the mix on 'solo safe.' If you then put something else on solo, you will always hear it along with the core of your mix.

Chapter 4

The Frequency Spectrum

No Light without Darkness

There is only one way to assess a frequency balance: by using your musical sensitivity. There is no absolute measure for such a balance. Without the low end as a counterpart, it's impossible to hear if something has too much high end. It can only be too much compared to something else.

In other words, a good frequency balance means that the various frequencies relate to each other in a musically meaningful way. There are no hard-and-fast rules on how much of a certain frequency is allowed. That's what makes mixing fun (and difficult): the optimal frequency balance is different in every mix. There are even multiple frequency balances that can work in the same song, depending on the instruments you place loudest in the mix.

Placing a nervous tambourine part loudly in the mix allows for less high in the entire frequency balance, because the jangling would probably attract too much negative attention. However, this changes when you reduce the volume of the tambourine and make the vocals louder. Your mix balance determines what frequency balance will work for the music, which is why it's a good idea to first set the best possible track levels before you start using EQ.

4.1 Association Is Key

The hardest thing about equalization is that your ears tend to get used to a certain timbre. Just listen to an album that sounds too dull at first. If you keep listening for half an hour, you'll slowly start to discover more detail and clarity in it, and after the fifth song you won't really notice the dullness anymore. But if you pause the album halfway and continue listening after ten minutes, you'll immediately notice the dullness again. Your first impression is essential to your ability to detect problems. There are two things you can do to not let a sound image with too much masking or an uneven frequency balance grow on you:

- Take plenty of breaks, and make sure to listen to a sound image that's different from your own mix during these breaks. The chirping of birds, the rustling of the wind, even the television news can 'reset' your ears.
- Your analysis of problematic frequencies should be quick and spot-on, so you can immediately make targeted interventions after you return from your break, before you get used to the sound image again.

Of course, the second point is the hardest to achieve—for it requires a lot of training—but you can speed up this process if you consciously start building a library of timbres and their related frequencies in your head. Over time, this library will allow you to connect a color, feeling or other

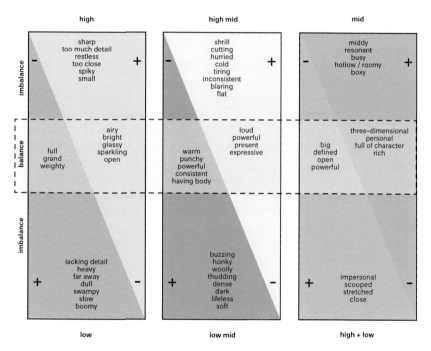

Figure 4.1 In order to assess a frequency balance, it can be useful to map your associations with different frequency ranges. This will give them a more concrete meaning for yourself. In this example, you can see three different frequency ranges set against their opposites, like yin-yang symbols for audio: if you have a lot on one side, there's bound to be less on the other. The diagram shows my own associations with these frequency ranges, both when they work well for the mix (in balance) and when they go too far into extremes (imbalance). As you can see, the same imbalance can have two causes: too much of one thing or too little of the other.

Everyone Hears Things Differently

I was reminded of the personal and cultural nature of sound associations when I gave a presentation at a Tanzanian music academy. The students all agreed that they associated low frequencies with coldness and high frequencies with a feeling of warmth, which are the exact opposite associations most Westerners have with these frequency ranges. After some questioning, I got the perfectly reasonable explanation that the low-end thumping implies unrest and danger, while the clear highs represent closeness and beauty. It was an extreme example of the fact that 'my high' is definitely not 'your high.' Sound associations are an indispensable tool for a mixing engineer, but they're also highly personal and therefore very hard to communicate.

association to a certain timbre, which will make this timbre or even a specific frequency easier to detect. Figure 4.1 gives you an example of a timbre library, but merely to illustrate the principle. It's no use to learn it by heart: a library can only be a good tool if it contains your own associations.

4.2 Masking

It hardly ever happens that your mix works perfectly without the need for EQ. But why is this? Assuming that all your source material sounds the way it was intended, that it was played with the right intention on good-sounding instruments, and that there's no coloration due to poor acoustics or microphone choices, there is only one thing left that can mess things up: masking.

Masking is the effect that can occur when two or more different sounds are played at the same time. As a result, you can't perceive all the sounds individually anymore: the dominant sound masks the other sounds. Properties that can make a sound dominant are, for example:

- relative loudness: when you have two sounds with similar sound characteristics, the loudest sound will mask the weaker one. The masking effect is much stronger when both sounds are coming from the same direction;
- a wide and dense frequency spectrum: sounds like snare drums have a noise-like tone and fill a wide frequency range with a lot of energy;

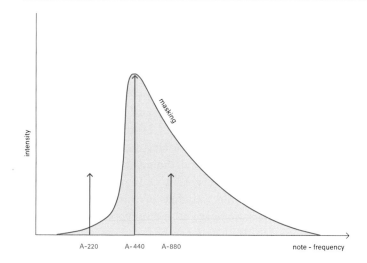

Figure 4.2 The principle of masking. The three arrows represent three sine tones, of which the two red ones are equally loud. The middle tone, the loudest one (A440), masks the sounds that fall within the gray area. Since loud sounds mask other sounds more in the frequency range above their own frequency, the low tone (A220) is still individually audible, while the high one (A880) isn't.

- pitch: it's more likely for low frequencies to mask high frequencies than the other way around (see Figure 4.2);
- duration: long-sustained sounds leave few openings for other instruments, unlike pointed sounds.

If you play two different sine tones at the same time and you keep amplifying the lower one, there is a clear moment at which it completely masks the higher tone, making it inaudible. In practice, masking is much more subtle, because sounds are complex combinations of various frequencies. Therefore, complete masking is something you won't often come across, but partial disappearance of sounds happens all the time. This isn't a problem in itself: a mix will sound boring and fragmented if one of the elements doesn't take the forefront every now and then, partially pushing the other elements away.

4.3 The Goal of EQ

When you make your initial mix balance, you create a hierarchy among the various elements, as described in Chapter 3. You examine which key elements are allowed to partially mask other, less important, elements. The next step is to control the masking: to determine for each sound which frequency areas can be masked, and which areas should remain audible at all costs. To get a knack for this, you must learn to assess the spectral function of a sound in an arrangement. Is it there to provide filling, depth, warmth, rhythm, melody, sparkle, space, bite, or something else? Your musical taste and perception play a major role in this, as you're constantly defining which elements of a song represent its energy, emotion and identity the most.

In many songs, the midrange in the vocals is important, but it might just be that the unique hook of a song is in the midrange of the bass guitar, or in the highs of the delay on the percussion, for example. And sometimes you find out that you don't need the highs of the guitars at all. In the end, you want to use EQ to highlight the frequency areas of

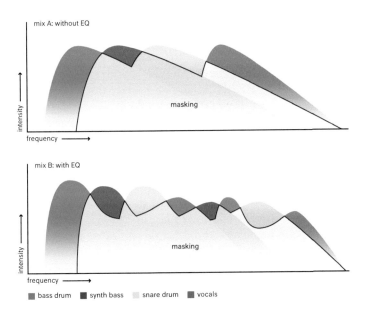

Figure 4.3 A schematic representation of the masking in a mix with four instruments, without and with EQ. Each color represents the frequency spectrum of an instrument. Without EQ, it's mostly the fundamental tones of the instruments that cut through the mix unhindered. This makes them audible, but you won't perceive a lot of detail. With EQ, you can make some higher frequency ranges that are important to the sound and musical function of an instrument (like the 'click' of a bass drum or the hammers of a piano) come out as well. To create space for these ranges, you also need to cut frequencies in the other instruments. In the end, the mix with EQ will translate better to different listening conditions than the mix without EQ. When all the elements are audible in the midrange, the low end can disappear on laptop speakers, but you'll still hear the bass.

the different sounds that are important to the music, and give them a place in the mix where they're not masked. The difference between the random masking in your initial mix balance and intended, EQ-controlled masking is visualized in Figure 4.3.

Research

A common beginner's mistake among mixers is making all the EQ adjustments to an instrument with its channel on solo. This way, you can hear the effects of your adjustments perfectly, but you learn nothing about their influence on masking in the mix. Still, solo listening can definitely be useful, because it can really help to map the various characteristics of a sound. By using parametric EQ to considerably cut or boost different frequency ranges of an instrument, you can find out what these frequencies mean to this sound. This way, you can create a mental 'frequency map' of the characteristics of a particular sound. For example, your research might tell you that the weight of an electric guitar is at 80 Hz, the punch at 100 Hz, the warmth at 150 Hz, the woolliness at 250 Hz, the tone at 500 Hz, the attack at 1 kHz, the harshness at 2.5 kHz, and the bite at 4 kHz. If you then turn solo off, and a characteristic is lacking in the mix as a whole (or if you hear too much of it), you'll know exactly in which frequency range to find it.

Perspective

Besides using EQ to control masking, you can do much more with it. Adding width and depth to your mix might not be the first thing you're thinking of (because panning, reverb and delay do this as well, of course), but the right EQ decisions can make the illusion even more powerful. Particularly the treatment of high frequencies can really benefit your mix in terms of perspective.

High frequencies won't easily cause masking problems, as they hardly get in the way of other sounds. Because of this, it can be tempting to make all sounds nice and clear and well defined: more high end generally sounds better. However, this results in a very strange sound image, in which everything is audible, but the perspective is gone. This is due to the fact that in a natural situation, the high frequencies from distant sound sources are damped by the air on their way to your listening position. The reflections of these sounds lose even more high frequencies, because they're absorbed by the porous materials in the listening room. So when you make everything equally bright in the mix, you take away an important mechanism to evoke an illusion of depth.

Equalizer Basics

A filter can be seen as a sieve that allows certain frequencies to pass through and blocks others. In analog filters, capacitors are often used as a sieve, as they have a progressively lower resistance for high frequencies (reactance). If you combine a capacitor with a resistor in a simple circuit, you can divide the frequency spectrum into high and low frequencies. This is called an RC filter: R for resistor, C for capacitor. The way the capacitor is connected in the circuit (in series or parallel) determines if the high or the low frequencies will reach the output.

The characteristics of the capacitor and resistor determine which frequency is the dividing line. This frequency is called the cutoff frequency, which is defined as the point where the filter attenuates the level by 3 dB. This way, you can use a resistor and a capacitor to build the two most basic filters: low-pass and high-pass filters. Plus you can put together a band-pass or band-stop filter by combining a high-pass and low-pass filter in series or parallel, respectively. The attenuation of these simple (first-order) RC filters increases by 6 dB per octave from the cutoff point, which gives them a very mild, natural sound.

EQ

When we talk about filters in everyday life, it's usually about high-pass and low-pass filters, and not about the filters found in the

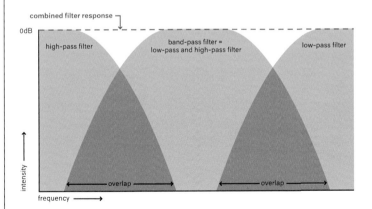

Figure 4.4 By combining high-pass and low-pass filters you can create a crossover. The middle frequency band can be seen as a band-pass filter.

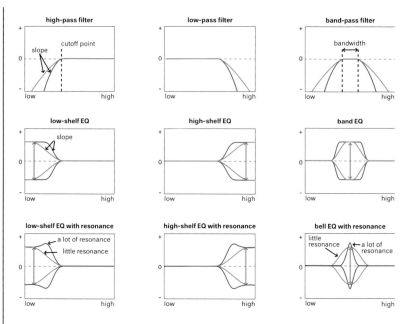

Figure 4.5 Common filter types.

average equalizer (which are often simply called EQ bands). The latter are a bit more subtle in their effect, in the sense that you can control the amount of amplification or attenuation of a particular filter band. Instead of increasingly cutting everything beyond the cutoff frequency, such a filter can also boost the midrange around 500 Hz by 2 dB, for example. The most flexible filters can be found in parametric equalizers, which often have three to six filter bands that can be adjusted freely in terms of their center frequency, bandwidth and amount of amplification or attenuation. You can see these as adjustable band-pass filters that you add to or subtract from the original audio at the desired ratio.

Many equalizers have shelf filters as well. These can be seen as high-pass or low-pass filters that you blend with the original audio. This way, you can boost or cut all frequencies above or below the cutoff point with an equal amount (see Figure 4.5).

Resonance

If you add an inductor to the circuit with the capacitor, you create a circuit with resonance. And if you don't use a resistor in this circuit to damp the resonance, you can even build an oscillator this way (as long as you feed the circuit with energy to resonate). You can also create resonance by adding an active amplifier to the circuit and supplying it with feedback. This creates a second-order filter element in the base, with a slope of 12 dB per octave. By varying the amount of feedback (or resistance in the passive circuit), you can influence the slope around the cutoff frequency of the filter. This is called the Q factor, or the amount of resonance. Besides an extra steep slope, very high Q factors (a lot of resonance) also result in a boost around the cutoff frequency (see Figure 4.6).

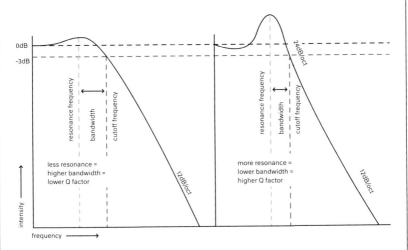

Figure 4.6 The effect of resonance on a second-order low-pass filter. The resonance in active filters determines their sound character to a great extent. The ringing sensation that these filters have can work great on a synth, but is usually not as welcome in an equalizer.

Interaction with Panning

There's a good reason why this book first discusses the subject of panning and then EQ, because the amount and type of EQ you need highly depends on the position of sounds relative to each other. For sounds that are heavily panned to one side it's often easier to attract attention, since they aren't masked so much by other sounds and don't suffer from crosstalk because they're only being reproduced by one speaker instead of two. This also means that they usually need less EQ to cut through the mix. So if you equalize first and then start panning, you'll have to readjust the EQ settings to the new masking scenario.

4.4 EQ Compass

What could be nicer than using EQ to make all the sounds as beautiful as possible? A bass drum with the depth of a thunderclap, guitars as sharp as knives, sparkling percussion, strings as wide as the sky, life-size vocals, and so on. Even if you're so amazingly skilled that you can make all these sound characteristics come out in your mix, this is usually not a good idea. It's better to choose a frequency emphasis for each sound that suits its musical role. For instance, if the tempo of the music is high, an overly deep foundation can make the mix slow and swampy. In that case, it's better to move the emphasis of the bass instruments a bit higher, even if this means more overlap—and therefore masking—with the other elements. Music that has to feel energetic and loud requires a powerful midrange, while sparkling highs and warm lows are less important. On the other hand, music that should be danceable requires more focus on the highs and lows of the beat, because if the midrange instruments attracted too much attention you would lose the hypnotizing rhythm. For intimate-sounding music, too much buildup in the upper midrange is a no-no, as this will make it sound agitated. In the best mixes, EQ is used to add definition, but also to give the mix a frequency emphasis that suits the music.

Dancing at the Center of Gravity

In order to be able to say something about the frequency balance, you have to start somewhere. Usually it's a good idea to define the low end of the mix first, and measure the rest against it. The lower limit of a song is the deepest frequency range on which it leans, and I call this the center of gravity. There can be frequencies below the center of gravity, but these are less important for the feel of the song. The center of gravity is the mechanism that makes your head bob up and down to the beat of the music. It has a very clear influence on your perception of the rhythm and tempo of a song. If the center of gravity is well chosen in terms of frequency and well measured in terms of emphasis, you'll be carried away by the rhythm of the music as a listener: you'll intuitively move along with the groove. On the other hand, a poorly chosen center of gravity can block this feeling. If it's too low, the rhythm will feel too sluggish to carry you away; if it's too high, it will feel restless and won't make you move either.

As a rule, a higher tempo requires a higher center of gravity to prevent the music from feeling slow. Note density is also important here: a bass that plays sixteenth notes can't have the same deep emphasis as a bass that plays quarter notes, for the same reason. Therefore, in heavy genres like metal or punk rock, the bass drum often sounds more like a 'click' than a 'boom.' But there are also genres that solve this problem in a different way: by adjusting the arrangement. In up-tempo electronic music

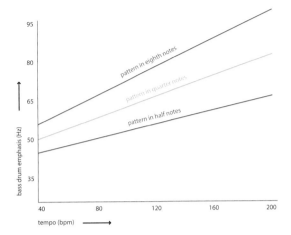

Figure 4.7 The relationship between the 'center of gravity' of a bass drum, the
tempo and the note density. The lines are approximations, and you
shouldn't see them as a tool to determine the emphasis of the bass drum
in your own mix, but as a demonstration of a principle. The higher the
note density and tempo, the higher the emphasis of the bass drum will
have to be for the rhythm to still work.

like drum and bass, the space above the low end is often kept relatively
free of instruments. As a result, the kick and bass are very 'exposed' in
the mix, so you can easily stay focused on the low end. With this kind of
music, it feels like there is a deep center of gravity, while the rhythm still
has a strong drive and a high tempo. Hip-hop capitalizes on the same
mechanism. Granted, the tempos are much lower, but there's much more
low in the frequency balance. The only way to make this work is by
reducing the low mids in the balance, and having a lot of high mids in
the vocals and snare drum to serve as a counterweight to the heavy low
end. So the way you shape the center of gravity of your mix is an interplay
of arrangement, tempo and note density. And of course personal taste,
although you'll find that it often matches the center of gravity that makes
your head move to the music. Therefore, the amount of low end in your
mix is also a great tool for manipulating your perception of timing.
Attenuating the low end automatically puts more emphasis on the light
parts of the measure and gives it a more up-tempo feel, while boosting
the low end emphasizes the heavy parts of the measure and makes the
mix more slow and sluggish.

Role Division

Another way to create a starting point for your frequency balance is to
begin with the most important element of the mix, and make it sound as

Don't Look

Many modern digital EQs contain a spectrum analyzer. Although this can be useful sometimes, it's often more of a distraction than an aid. The problem lies in the fact that the analyzer doesn't know the role of an instrument in the mix. Maybe those sharp peaks in the spectrum are actually the character that the sound should give to the mix. Analyzers will easily trigger an aversion to outliers, and lead you to flatten the spectrum. Besides this, frequency components that cause a lot of disturbance—and that you would want to cut with EQ—might seem completely insignificant to the analyzer. Think, for example, of frequency components caused by distortion. These are relatively weak and therefore barely visible, but clearly audible. So it's better to trust your ears than your eyes.

nice as possible in itself. Then you try to make as few concessions to this element as necessary, by adjusting the other instruments to it. This way, conflicts are very easy to resolve. Is the voice the most important element? Then some of the piano's mids will have to be sacrificed, if those two are clashing. Is it the kick drum that drives the music? Then you should remove some more lows from the bass line. With this method, you almost automatically end up with a mix in which the priorities are clear, because it creates a distinction between the foreground and background. In the beginning it will help to exaggerate your EQ actions a bit. Such a rough draft is more robust and provides a good foundation to build on. This works better than a tentative sketch that has to be adjusted every time you make a change to the mix.

The role of an instrument is not always immediately clear, especially in the context of an arrangement. Many instruments perform more than one function: they play a rhythm, form chords and add a certain sound dimension (movement, texture, space) to the music as a whole. But which of these functions are important for your mix—and which are of minor importance—you'll have to find out for yourself. Often it's obvious and you can just follow your intuition, but sometimes the choice also depends on what the other elements are failing to do, or on the role that's still vacant in the mix. Is the drum performance not completely convincing? Maybe it will fall into place if you emphasize the rhythm of the guitars. Is your mix chock-full of vocals? Then a doubled version of the lead won't add much in terms of tone, but it will in terms of space. Of course, the distinction between different functions isn't always so clear-cut, but you can definitely emphasize or weaken functions.

Tuning

The pursuit to make the different parts fit together can best be started in the context of the mix. I call this part the 'tuning' of the instruments; not in terms of pitch of course, but in terms of frequency emphasis. This method is ideal for additional parts that don't necessarily have to be clearly recognizable, but that should mainly strengthen the core of the mix. You take an equalizer with a wide bell boost, of which you adjust the frequency with your left hand, and you control the fader of the channel you're tuning with your right hand. This right hand is very important, because every frequency adjustment will also cause a balance difference in the mix, which you immediately have to compensate for with the fader to assess whether a particular frequency emphasis works.

A digital EQ with auto gain (automatic compensation for the level difference caused by the EQ) would make this process easier, but the position of an element in the mix can still change depending on the frequency range you're emphasizing. So you'll always need the fader. If you get the hang of this system, you'll easily hear which aspects of a certain sound are important, and which actually add very little or even cause problems in the mix. At first, try not to pay too much attention to the clarity of the mix as a whole. Listen if you can hear the desired musical aspect emerging from the mix: growl, tone, firmness, clarity, sharpness, size or rhythm. But also listen how it affects the feel of the

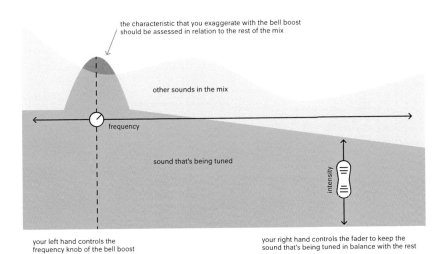

Figure 4.8 By sweeping the center frequency of a wide bell boost, you can roughly adjust the frequency emphasis of a sound, to hear where it could find its place within the mix. With the fader, you can immediately adjust the mix balance to test your idea. Start by using a wide bell: this will sound most natural, and it's a good choice for setting up a first rough draft.

Interaction with Compressors

Compressors can have a significant impact on the timbre of a sound. The frequency range that triggers the compressor the most determines the role the sound will play in the mix. For example, you can use EQ to give a bass guitar—which already seems to have a decent amount of low end—even more low frequencies, and then use compression to even it out. This creates a deep bass with very consistent lows. But you can also remove some of the lows before you run the bass through the compressor, which will make the compressor emphasize the 'honky' midrange of the bass.

After the compressor, you can use another EQ to fine-tune the sound, for example by adding some deep lows again. This method—

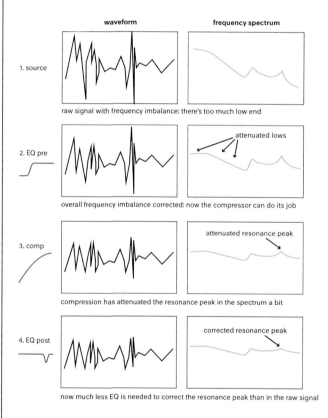

Figure 4.9 If you use EQ and compression in the right order, you'll need less of both to achieve your results—so their drawbacks won't be as much of a problem either.

using EQ to exaggerate a feature, holding this feature in place with compression, and then dipping any resulting surplus of certain frequencies—often works very well. If you first 'tune' the sound in the mix according to the method I just described, the compressor will never be excessively triggered by a role the instrument doesn't play in the mix. As a result, you won't need as much compression, and the compression you use is more effective. When it comes to minor corrections, it's better to make them after the use of a compressor, because otherwise your interventions will influence the character of the compression, which you were already satisfied with.

For example, after you remove some proximity effect from the vocals with a low shelf and add some bite with a broad boost around 5 kHz, you first add compression and then refine it by dipping the annoying narrow resonance at 2.6 kHz a bit. Because the compressor has stabilized the sound somewhat, small corrections have a much bigger impact than when the dynamics were still all over the place (see Figure 4.9).

song and how the other elements come across. When you find the frequency range that's responsible for the desired characteristic, you can fine-tune the gain to make the sound blend well with the rest of the mix. If the EQ changes the character of the sound too much, and if the desired character does come out when you listen to the element on solo, you should try to cut some of the corresponding frequencies in the surrounding mix elements.

More High End or Less Low End?

Locating problems that get in the way of a good frequency balance is a web of contradictions. All the frequencies that you boost or cut affect your perception of all the other frequencies. In that way, it's just like looking for the right mix balance, but with more possibilities. For example, in a mix that turns out to be sharp and chaotic, you can use narrow filters to reduce the disruptive resonances of metal percussion and distorted instruments. Often you will find (after using five or more EQ bands) that the problems won't go away—and worse, that you're losing a lot of good things because of your interventions. A problem like this can also be solved in a different way: as long as the low end is solid enough—which will make for a steady and calm listening experience—

to offset the sharp chaos, the latter will automatically be less apparent. In more gritty and raw-sounding genres, this is the go-to mechanism for keeping the music rough but listenable. As long as there's enough in the music for the listener to hold on to, the noise won't distract too much, and it will even help to achieve the desired atmosphere. This is not always easy, because in the music that's the hardest to mix (punk and metal, for example), the center of gravity is high up the spectrum due to high tempos and note densities, but there's also a lot of aggression in the high midrange caused by loads of distortion. This makes it quite a puzzle for the mixer.

Before you start going down a long road of problem solving, it's a good idea to quickly determine the location of the problem. A boost or cut with a simple shelf filter can instantly give you more insight. If removing lows immediately makes the mix more defined, a cut in that area is probably the best way to go. But if this turns your mix into a pile of mush, you probably need to add more highs for extra definition. It also works the other way around, because if a high boost makes the mix sound unnatural and sharp, it's probably a clash in the low end (or midrange) that's causing the problem. Removing highs is always difficult, as it rarely appears better than what you started with: it will mostly sound dull in comparison. A short break can sometimes help to figure out if removing highs is the solution. Conversely, adding highs is an almost addictive way of giving your mix extra definition. But it's often too good to be true: it won't solve the clashes in the midrange, plus it makes the mix flat and tiring to listen to. The contrast between different instruments in terms of the amount of high end is an important mechanism for creating depth in your mix, so if there's too much high end, this depth is lost. One way to avoid falling into this trap is to give yourself the artificial restriction that you can't boost any highs, but only cut other frequencies. This way, you'll automatically tackle the root of the problems, save for the rare case in which something is really too dull.

The less EQ you need, the better. Instruments sound fuller and more natural if you don't process them too much. But to make this possible, everything has to work out well: the performance, recording, arrangement, and your mix balance of course.

Limitations

Sometimes you have a sound in mind, like an electric guitar as thick as a wall, and you're trying to equalize the guitar you're working on in that direction. But if—after you've been turning knobs for half an hour—it turns out you're using eight EQ bands and it still doesn't sound the way you wanted, it's time to look for other solutions. An equalizer can only boost or cut frequencies that are already present in the sound. Chances are that the guitar you're mixing is played in a different key than the guitar you had in mind. Changing the key is usually not an option in mixing, so you'll have to come up with something else. There are a couple of tricks you can use to generate extra frequencies, instead of using EQ to boost existing frequencies. For example, you can use a sub-bass generator for bass lines or percussion parts that sound too light. Such a processor generates an octave (or two) below the input signal, which can add a lot of extra depth to the sound. This generally works well for monophonic sounds, but for the guitar I just mentioned it's probably a bit too much of a good thing. Of course, 'manual' pitch shifting is also possible, and percussion can easily be supplemented with a sample that does have the frequency content you want.

If an instrument doesn't sound big enough and a low boost or high cut makes your mix muddy and undefined, it might very well be that you actually need to cut some lows and add more space (reverb or delay). And if you can't feel the low end in terms of punch, but the mix still sounds woolly, there are probably too many sounds with sustain in the low range. In this case, it might help to use less compression, or to shorten the low-frequency sustain of some elements with transient designers or expanders. You can read more about this in section 11.3.

Besides this, overtones are very important to make a sound audible in the mix. Sometimes a sound doesn't have enough over-tones of itself, for example drumheads that are badly tuned or overly muffled, or old strings on a (bass) guitar. In those cases you can equalize all you want, but it won't give you any overtones. What it will do is make annoying noises unnaturally loud. So it's better to generate overtones yourself, for example through har-monic distortion or with an exciter. These kinds of processors also work well to make sure that low-frequency sounds are still audible on small speakers.

Chapter 5

Dynamics
Without Soft There Is No Loud

In music, the concept of dynamics refers to the difference between loud and soft. These differences propel the music. They provide tension, impact and release, reflection, melancholy or calm before the storm. At concerts, dynamics play a huge role, whether it's dance or classical music. This is only possible because the circumstances lend themselves to it. It's pretty quiet in a concert hall. In a room like that, a symphony orchestra can produce differences of up to 70 dB without the music becoming inaudible. Things can get a lot noisier during a club night or pop concert, but powerful sound systems make up for it. They allow for large level differences without anything getting lost in the murmur of the crowd.

This changes when you record one of these concerts. You won't be listening to the recording under the same circumstances as the original performance. To keep your neighbors happy, you won't dial up the volume of your living-room sound system to concert levels. And when you're on the go listening through your earbuds, the level of the background noise compared to the music is a lot higher than it was in the concert hall. In other words, if you want to create a recording that a large audience can enjoy in many different situations, a smaller dynamic range will be available to you than during the original concert. The solution seems easy. Run your recording through a compressor and there you go: the dynamic range has been reduced. But of course, this will also diminish the musical function of the original dynamics, making the recording sound dull and flat. The trick is to manipulate the dynamics in such a way that they become smaller, without losing musical impact. You want to make the dynamics seem much larger than they really are. So how does this illusion work? The key is in the development of dynamics over time.

5.1 Dynamics versus Time

The word 'dynamics' can be used for the difference between two notes, but also between two sections of a song. In a piece of music, dynamics occur at a variety of timescales (see Figure 5.1), so if you want to manipulate dynamics, it's a good idea to consider at which timescale you want to do this. And even more so, to decide which timescales you want to leave alone because they're important for the musical development and impact. For example, if you want to reduce the level differences between a couple of notes with a compressor, it should work so fast that it attenuates the loud notes and leaves the weak ones alone. But besides decreasing the short dynamics between individual notes, compression does the same with the longer dynamics that exist between measures, phrases and sections. And these long dynamics are exactly what gives the music tension and development. Not surprisingly, mix engineers who specialize in ultra-compact pop music first use compressors to make sounds more dense, and then restore the overarching dynamics through fader automation. Adjusting the dynamics on a short timescale also affects the dynamics that occur at larger timescales. If you don't do anything

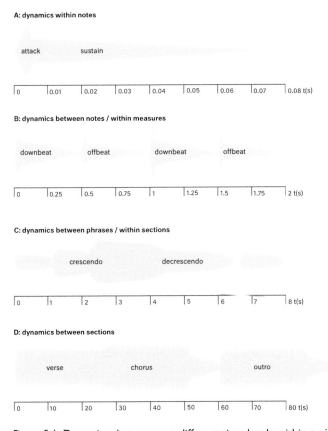

Figure 5.1 Dynamics that occur at different time levels within a piece of music.

Compressor Basics

A compressor is nothing more than an automatic fader that turns the volume down when it exceeds a certain level. It's mostly the detection circuit that determines the behavior and sonic character of a model. This circuit controls the 'fader' based on the characteristics of the incoming audio. The level at which a compressor is engaged is called the threshold. Sometimes this is a fixed point (hard knee), but often the compressor already starts working at a level below the threshold, increasing the amount of compression as the sound level gets higher (soft knee). The attack time controls how fast a compressor reacts to a signal that exceeds the threshold, and the release time is the speed at which the signal returns to the original level after compression. The ratio determines how severely a compressor intervenes when the threshold is exceeded. A ratio of 2:1 means that a note that exceeds the threshold by 6 dB is attenuated by 3 dB. A ratio of ∞:1 (combined with a very short attack time) means limiting: no signal is allowed to exceed the threshold. Finally, most compressors have a make-up gain control to compensate the loss of level they cause. In section 17.6 you'll read more about how compressors work.

about this, compression can easily ruin the impact of your mix. The opposite is not true, because if you use a very slow-reacting compressor to bring a few vocal lines closer together, you'll hardly change the dynamics between individual notes.

Categorizing dynamics in timescales makes it easier to understand what exactly you can influence with the attack and release settings on your compressor. For example, you often hear people talk about adding punch with a compressor. This might seem counterintuitive, because a compressor turns loud passages down, which is literally the opposite of adding punch. The secret lies in the fact that the first onset of a note, the transient or attack, is most important for the punch. So if you give a compressor a relatively slow attack time, it will influence the ratio between the onset and the sustain of a note. It allows the transient to pass before it starts to turn down the sustain. As a result, the sound will have proportionally more attack: more punch.

You then use the release time to determine the character of the sustain of the notes. An extremely short release time will make the effect of the compression too small, so it won't add a lot of extra punch. A short release time will bring the last part of a note's sustain back to its original level. Within the course of a note, you can hear the compressor bring the level down and back up again: a distinctive sound that's often described as 'pumping.' An average release time will make sure the compressor's level is up just in time to let the onset of the next note pass before it brings the level back down. If you want the compressor's effect to be as transparent as possible, this is the best setting. The ideal value of this setting depends on the note density and tempo (see Figure 5.2). With a long release time, the compressor is always too late to let the next note pass, as it's still busy compressing the previous one. Such a release time gets in the way of the music's rhythm and ruins the groove. However, this is only the case if you're using the compressor to add punch. If you don't intend to change the sound of the onsets of individual notes, you can let the compressor respond to the average energy of entire musical passages. In that case, a long release time won't be a problem at all. For example, in vocals the dynamic arcs can get quite long, so if you want to manipulate these with a compressor, it's a good thing to have a release time of a

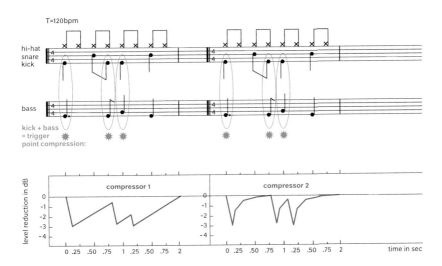

Figure 5.2 Two compressors, both with a slow attack and release time, manipulate the same measure of music. The threshold is set in such a way that every coinciding kick and bass note triggers the compression. Both compressors have the same slow attack time, so they let the musical center of gravity pass before they intervene. The sound difference is in the shape of the release curve. Compressor 1 has a very gradual release that can sound sluggish, and in this case it means that only the first beat of the measure and the fourth eighth note are emphasized. Compressor 2 initially has a relatively fast release that gradually becomes slower. Sonically, this can be perceived as punchy, and in this case all centers of gravity are emphasized, also the one on the second beat.

couple of notes long. This will prevent you from evening out the course of the arc, as the compressor will 'see' the arc as one unit.

5.2 Technical and Musical Dynamics

In theory, you could perform all the compressor's functions yourself through the automation of fader movements. The downside is that you'll need to reserve a few extra days for some mixes, but there's also one big upside, because you'll be using the most musical, most advanced detector to control the compression: your own ears. Setting a compressor to make it react in a musical way isn't easy. In this case, 'musical' means that it reacts to dynamics just like the average listener's auditory system does. The auditory system has a number of unique features, because besides varying by frequency, our perception of loudness is time-dependent as well. And of course this last aspect is very important when setting a compressor.

If you stand next to a decibel meter and you clap your hands loudly, the meter can easily show 110 dBSPL. But if you then try to match this sound level by screaming, you really have to make an effort. The primal scream that finally manages to reach 110 dB feels much louder than the handclap you started with. This is because the length of a sound plays a major role in our perception of loudness: we perceive short bursts as less loud than long sounds. So if you want a compressor to 'hear' things like we do, you'll have to make it less sensitive to short peaks. Fortunately, this is easy to achieve by setting a long attack time, but to many beginners this seems counterintuitive. This is because they think compressors are meant to cut peaks, so you can make sounds louder in the mix without the stereo bus clipping. Although you can definitely use compressors for this purpose, in the end this approach will result in weak-sounding mixes.

It's very important to realize that compressors are meant to change musical proportions, and that this has nothing to do with peak levels. Musical proportions are in the 'body' of the notes, and hardly in the transients. When you start cutting transient levels, you're changing the technical dynamics: the peak level indicated by the meter. A good limiter will do this virtually inaudibly (within a limited range), while it doesn't change the musical proportions at all. This is exactly why a limiter is so uninteresting as a mixing tool: it doesn't affect the proportions, especially if it has adaptive release. So if you're not aiming for effects like boosting the reflections in room microphone tracks or solving specific problems with excessively sharp transients, you're better off with a compressor that leaves the peaks alone. You want to control the musical, not the technical dynamics.

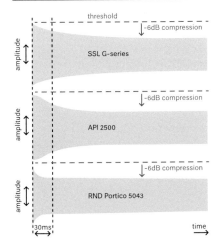

Figure 5.3 Three compressors are set at ratio 3:1, with an attack of 30 ms
(according to the front panel), and 6 dB gain reduction. After the onset
of the test signal, the SSL and API react as expected, and after 30 ms
most of the level reduction is over. The RND seems ten times as fast,
but it also takes 30 ms for it to reach its final level reduction. This is
not a subtle difference in sound! As you can see, the compressors
all have different fade-out curves right after the attack. This is an
important part of their sound.

In the old days, it was a matter of course that a compressor didn't
respond to extremely short peaks, as the electronics simply couldn't keep
up. However, in this age of plugins it has become a conscious choice to
set the attack time long enough. Keep in mind that there's no fixed
standard for specifying the attack time, so always use your ears and don't
look at the numbers (see Figure 5.3). When you turn the attack knob
from the shortest to the longest setting, you'll automatically hear the
character of the compression change. From overeager and stifling to a
more open sound, and then to the point where the dynamics don't really
change anymore because the compressor reacts too late.

5.3 Using Compressors

Compression has a reputation of being a miracle cure that can instantly
give music more impact, loudness, consistency, tension and who knows
what else. But if you're not careful, it will make the music small, dull,
boring and one-dimensional instead. The difference between this sonic
heaven and hell is often in minuscule setting changes. It's enough to make
you nervous: one wrong turn of a compressor knob and the sound
collapses like jelly. You get confused, start turning the knobs even more,

and slowly you forget how good the source instrument initially sounded, and how little you wanted to change it at first. Your reference is sinking further and further, and it's not until the next day that you notice how much you've been stifling the mix with your well-intended interventions.

Hear What You're Doing

The only way to avoid straying from the direction you wanted to go with your mix is to thoroughly compare if your manipulations actually make things better. Based on this comparison, you decide if the pros outweigh the cons (every manipulation has its drawbacks!) and if you need to refine anything. But making a good comparison is easier said than done. Just conduct the following experiment: tell an unsuspecting listener about a new processor you've used to give your latest mix more impact. To demonstrate its quality, you prepare the same mix on two different tracks, titled 'A' and 'B.' You don't say that the only difference between the two is that you've made mix B 0.3 dB louder than A. Then you proudly switch back and forth between the two mixes, while you ask the listener to describe the differences. Chances are that the test subject will hear non-existent differences in dynamics, frequency spectrum, depth and stereo image. Due to the slight loudness difference between the two mixes, you can't objectively compare them anymore.

Because of this, the most important thing when setting a compressor is using the make-up gain parameter. This will allow you to compensate for the loss in loudness caused by the compressor, so you'll keep the audible loudness difference to a minimum when you switch the compressor in and out of the signal path. Still, this is easier said than done, because a compressor—unlike an equalizer, for example—doesn't cause a static difference in loudness, but changes it with time. Since it's not possible to constantly compensate for the varying level change—which would ultimately undo the compression, should you succeed—it's best to set the make-up gain by gut feeling. Try to give the sound the same place in the mix, whether the compressor is on or off. If you have doubts about the amount of make-up gain, it's better to use a bit less. This way, you're less likely to think the compressor has a positive effect, and it makes you work harder to really find the optimal setting.

The final hurdle for the objectivity of our before/after comparison is time-related as well. Because the effect of a compressor is so variable, it's hard to assess its impact by switching it in and out during playback. The effect of the compressor on the line right after you turn it on isn't necessarily the same as it would have been on the preceding line. You can't predict the effect of a compressor on the entire song based on an instant comparison, like you can with the effect of an equalizer. Therefore, it's important to listen to a section of the song that's representative of its

How Much Compression?

Compression can help to keep the lows in check and make sure the mix still sounds full at a low listening level. But if you go too far, the low end will lose all its impact, because getting rid of all the peaks means throwing out the punch as well. You can easily check how much compression you need: at a high level, there should be enough punch, but when you turn the level down you should still be able to hear the lows. If you don't feel enough punch, you're using too much compression, or your attack times are too fast and your ratios too high. But if the low end disappears when you turn the music down, you'll need more compression, possibly with faster attack times and higher ratios.

overall dynamics, both with and without compression. Only then you can assess how the compressor affects your perception of the dynamics, and whether this is an improvement.

There are compressors that automatically adjust their make-up gain to the amount of compression, which would facilitate equal-loudness comparisons. However, I've never heard this work well in practice, and although it can provide a good starting point, you can't trust it completely. Your own ears are a much more reliable tool for estimating loudness differences. Besides this, the meters that display the amount of compression (gain reduction) aren't the best guides to base the amount of make-up gain on. These meters can give you a rough indication of what's going on in terms of compression, but they hardly tell you anything about the sound. In fast compressors that use VU meters (or a digital simulation of these) to display gain reduction, peaks can already be significantly compressed before there's any deflection on the meter. This is because the meter reacts too slowly to display the compression. Conversely, the gain reduction meters in digital limiters (which only block peaks) can react very strongly, while the peaks hardly mean anything in terms of perceived loudness. So remember to only trust your ears!

Zooming In

Compression is often used to solve a specific problem. For example, a vocal recording contains loud bursts that suddenly stick out sharply, which makes it hard to position the vocals in the mix. You can solve this by setting the compressor in such a way that it mainly blocks these annoying peaks and leaves the rest alone. This can be tricky, for example when the vocalist sings a low line in the verse that contains a relatively large amount of energy, but that doesn't go off the map in terms of dynamics. The compressor doesn't see any difference between a low note that feels balanced, and an equally intense high note that you perceive as piercing and too loud. Because of this, you sometimes need to lend the compressor a hand, by exaggerating the problem it has to solve.

Many compressors have a built-in sidechain filter, or the option of patching a processor of choice in the sidechain. This way, you can filter the signal used by the compressor for its detection (not the signal it's processing). In the example of the vocal recording, you could boost the 2 kHz range in the sidechain of the compressor. This will make the compressor perceive the piercing high notes in the vocals as much louder than they really are, so it will apply more compression to these bursts. With the filter, you increase the level difference between the peaks and the rest of the signal, which makes it easier to only apply compression to the bursts and not to the low vocal line in the verse. This method can also be used to turn a compressor into a de-esser. If there's no need for very extreme corrections, such a (broadband) de-esser can sound more transparent than a de-esser that uses dynamic filters.

Sidechain filters can also be used to make a compressor respond to loudness more like our ears do. Not a bad idea, since you usually need a compressor to balance sounds based on human auditory perception, not on how they look on the meter. At an average listening level, our ears are less sensitive to low frequencies than to midrange frequencies. With a sidechain filter, you can give a compressor the same frequency dependence. Of course, it won't work exactly the same way—as the hearing curve depends on the sound pressure—but its overall shape is still a usable reference. Many compressors have a high-pass filter in their sidechain, which can be very useful here. Thanks to this filter, the low frequencies that sound comparatively weaker to our ears will also have a relatively smaller share of the overall sound level detected by the compressor (see Figure 5.4).

Serial Compression

Compressors come in many flavors. In section 17.6, I'll discuss many different compressor types and their strengths and weaknesses. In practice, you'll often need to solve complex problems that require more than just the strengths of one specific compressor. What you want to avoid at all times is using a compressor outside its 'comfort zone.' Outside that zone, the compressor will have too many negative effects on the sound, the most common of which are 'pumping,' 'clogging' and distorting.

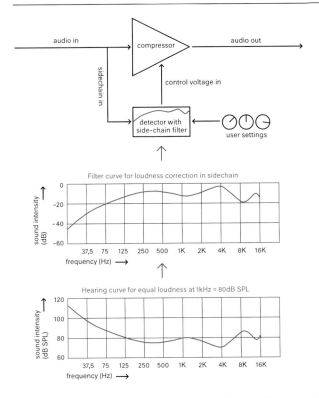

Figure 5.4 From the ear's response you can derive a filter curve for the sidechain of a compressor, which will make it respond to loudness in a way similar to how we perceive it. Sometimes it can be useful to not manipulate the sidechain signal with EQ, but the audio signal itself. For example, if a vocal recording contains loud popping sounds, it's better to filter off some lows before you run it through a compressor. If you do it the other way around, every single pop will push the compressor into maximum gain reduction, because it detects a high peak. In other cases, it's recommended to first compress and then EQ, especially if you've already found the right dynamics for a sound with the compressor, but you still want to refine its place in the mix with EQ. In that case you want to avoid changing the dynamics again with every single EQ adjustment you make.

Because of this, it's sometimes better to break complex problems down into sub-problems and start tackling them one by one. An example: a recording of a plucked double bass contains large differences in dynamics between notes, plus there are extreme peaks on some of the plucks. Instead of using one compressor to solve both problems, you let two different compressors work in their own comfort zone. First, one compressor takes care of the peaks with a fast attack, fast release and high ratio. Now the peaks won't affect the second compressor anymore, which balances

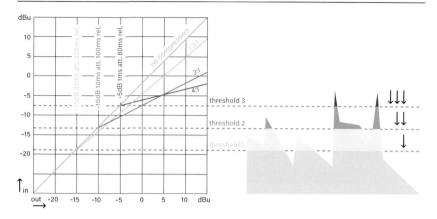

Figure 5.5 A soft knee response made up of three serial compressors with increasing ratios and timing.

the average energy between the notes with a lower threshold and a much longer release time.

Mastering compressors by Weiss and Massenburg use this principle to tackle complex signals transparently. They consist of three compression sections in series, of which the first one has the lowest threshold, lowest ratio and slowest timing. The next two both take it a step further with a higher ratio and faster timing, but with increasingly high thresholds. This system can be seen as soft knee with a complex response (see Figure 5.5).

Parallel Compression

Another way to combine the advantages of different compressors is to connect them in parallel. For example, you can use aux-sends to run a vocal recording through three compressors: a fast compressor to make the voice sound clear and up-front, an optical compressor to make it warm and smooth, and an aggressive compressor as a special effect. The three compressed sounds return to the mix on three different channels, which you can then balance any way you want. A bit more to the front? More of compressor number one. Too flat? More of number two. Too polite? Boost the third one. This is a very intuitive process, and you can even use automation to adjust the compressor balance to the different sections of a song! Keep in mind that if you work digitally, the latency of the compressors must be properly corrected. If not, a parallel setup like this can cause some pretty annoying phase problems (see sections 7.1 and 13.3).

Of course, besides the compressed signals, you can also add the original signal to the mix. Some compressors come with a wet/dry

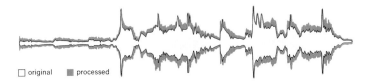

□ original   ■ processed

Figure 5.6 The effect of parallel compression: the peaks of the processed signal (orange) are still virtually identical to those of the original (yellow), but the information in between is boosted.

control, but you can easily use an aux-send for this purpose as well. This is what is usually meant by parallel compression, and it's the ideal way to make certain details of an instrument come out without making it sound flat. For example, some compressors won't really sound interesting unless you 'hit' them pretty hard, and to prevent the sound from becoming too flat, you just mix in a decent amount of dry signal again. The effect of this is that the peaks in the original signal remain largely intact, while the signal unaffected by the compressor is boosted in terms of intensity (see Figure 5.6).

Making Connections

As a rule, individual instruments or samples will sound more connected if they have a joint influence on the sound image. Shared acoustics can act as a binding agent, but bus compression can contribute to this as well. Bus compression means sending a number of instruments to the same bus and putting a compressor on it. This can be the entire mix, or a select group of instruments that have to blend together more. Think of all the individual drum microphones, the drums and the bass, all legato parts, all parts that play on the beat, or any other combination.

If an instrument is relatively loud, it affects the overall compression, and therefore your perception of the softer elements as well. This makes for a much more exciting and consistent mix, in which everything affects everything else. Because of this, compression on groups of instruments is just as useful as compression on individual elements. The rule of thumb that usually works is that problems should be fixed by focusing on individual elements, but if you want to create consistency and compactness, it's often better to do this in groups. Chapter 12 will elaborate on the use of bus compression.

5.4 Reducing Masking with Compression

When two instruments play together, they sound different than when you hear them separately. Parts of one sound drown out parts of the other

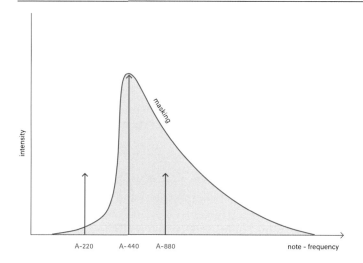

Figure 5.7 You'll remember this figure from the previous chapter: a loud tone creates a mask, making the other tones within this mask inaudible. In this chapter, the same problem is solved with compression instead of EQ.

sound, and vice versa. This simultaneous masking takes place as soon as a relatively weak part of one sound falls within the so-called mask of a relatively loud sound (see Figure 5.7). But it goes even further than that, because to a lesser extent, the masking already occurs right before (about 20 ms) and after (about 200 ms) the louder sound is played. This is called temporal masking. The closer together two sounds are in terms of loudness, the less they mask each other. And the more the sounds vary in terms of frequency composition and localization, the less they clash. Therefore, it rarely happens that a sound is completely inaudible in the mix, but in certain frequency ranges it can be pushed away entirely. Especially broadband sounds (which cover the full width of the frequency spectrum) and sounds that play long notes or whose notes trail (and therefore have a lot of overlap with other sounds) tend to drown out other sounds. Arena-sized drums and distorted guitars or synthesizers are notorious culprits when it comes to this. If all these instruments didn't play at exactly the same time, things would be a lot easier. But rock is not the same as ska, so you'll have to think of another way to solve this problem.

Dividing the Beat

Drums in particular are often played at the same time as other instruments, potentially causing them to mask a lot of each other's detail. Simultaneous masking like this can be reduced by making these sounds less simultaneous. Of course, you don't do this by shifting the drums and the other instruments away from each other, so they don't play at the same time anymore. But what you can do is divide a beat in several time zones, each dominated by a different instrument. For example, the first zone can be dominated by the attack of the drum sound, after which the guitars and then the bass take over, followed by the sustain of the drums, until finally the sustain of the guitar reappears: all of this within the 50 to 100 ms of a beat.

The nice thing about this concept is that you're less dependent on instruments that continuously dominate certain frequency zones, which happens if you use a lot of EQ to give sounds a particular place in the mix. Thanks to the different time zones, instruments that occupy the same frequency range can become more interlocked. This will make your mix much more robust when it's translated to other speaker systems, because in the midrange that's so important for this translation, all the instruments can now have a dominating presence, just not at the exact same time.

A Compressor per Place in the Beat

A system of time zones per musical beat is easy to explain, but how do you get something like this to work? The trick is to have compressors attenuate the parts of a sound that you want to reserve for other sounds. Take the drums, for instance. You want to emphasize the initial attack, and a bit later some of the sustain. By using a compressor with a slow attack and an average release, the part of the drums right after the first hit is attenuated. This leaves some room in the mix, mainly for the electric guitar. Usually you don't even have to manipulate the electric guitar for this, because its transients are already a lot slower (and less sharp) than those of the drums. But if you compress the guitar with a slow attack and release and a very low ratio, this will emphasize the guitar's sustain, so it can fill the gap after the drums have faded away. Next, you can run the bass through the same compressor as the drums, so it will be most strongly audible when the kick drum doesn't play. It's thanks to these mechanisms that compression is such a major factor in your perception of rhythm and such a useful tool for giving sounds a place in the whole.

You might think this type of compression affects the drums too much or makes them sound unnatural, but surprisingly this is not the case. Especially if you use the compressed signal parallel to the original, it can sound very natural and work like a zoom-in function. In the mix, the drums will sound more like they do when you listen to them on solo without compression. Why is this? Due to the Haas effect you don't

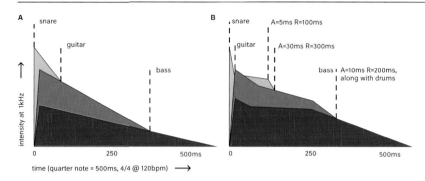

Figure 5.8 Compression can help with the temporal redistribution (envelope shaping) of different parts playing the same beat (A). This way, they will give each other more space without the need for EQ (B).

perceive every reflection as an individual sound source (you can read more about this in section 7.2), and as a result the time range just after the first hit of the drum is not so important for your perception of the drum's sound. In a way, your ear combines it with the first hit. It's not until much later that you perceive the acoustics around the drum again, which determine a great deal of the sound. So the middle time range of a drum hit might not contribute much to your perception of the drum, but it still takes up space in your mix. It's a relatively loud and long-lasting part of the drum hit, so it will get in the way of other instruments playing at the same time. You may not notice anything about the drum sound itself, but your mix isn't working out the way you wanted: all the energy and power seems to be blocked because you're wasting too much space. Compression can help to attenuate the part of the drums that doesn't contribute much to their size and power, but that does take up a lot of space. As a result, your perception of the drums in a busy mix can be more similar to how they sounded in the recording room.

Too Flat?

The drawback of using compression as an envelope shaper is that it affects the dynamics of the sound. A popular way to keep this effect within limits is by applying heavy compression to some drum microphones while leaving others unprocessed, so you still have some dynamics left. You could see this as a form of parallel compression. However, even this can be too much sometimes. When sounds already have a lot of density—like distorted guitars—the tiniest bit of compression can make them lose their definition and squash them into a tiring mush of sound. In that case, it's better to manipulate the envelope in different ways. For example, recording the guitars on an analog recorder with a low tape speed will

Denser Music Requires More Compression

The busier a mix gets, the less room you'll have for dynamics. When many instruments are playing at the same time, there's no room for one of these instruments to play much softer than the rest and still be audible. And on the upper half of the dynamics scale, there's not much space either: if one instrument suddenly gets very loud, it can get in the way of five other instruments and make a grand arrangement sound small. The higher the density of the notes, the higher the tempo, and the more instruments there are in a mix, the more compression you need to make them work together. Sounds can't get too loud or too soft if you want to maintain a working balance.

smooth the peaks in a way that won't make the guitars as flat and swampy as compression does. And a transient designer (which can change the level of the attack and sustain part of a sound, independently of the signal level) will serve this purpose as well.

Too Good?

In music that was made on a grid, there's a risk that the parts are so perfectly aligned that their individual character is barely audible anymore. Transients are the biggest giveaways of sound sources. You can see them as tiny attention grabbers that briefly announce a new instrument. If the transients of simultaneously played instruments are easy to distinguish from each other, the instruments themselves are likely to be distinguishable as well. You can create distinctions through the use of panning and timbre, but if the transients all occur at the exact same moment, it will be very hard to tell them apart.

In this case, the compression method I previously discussed can't make enough of a difference. With music played by humans you won't have this problem, because even though there are insanely great musicians out there, they're definitely not robots. Their interpretation of the part they play—by delaying or advancing certain notes—can even give it a comfortable place in the overall groove. However, with loops or other quantized material this doesn't happen automatically, and sometimes it can be worthwhile to move the timing of the parts around a bit (see Figure 5.9). Running an instrument through a distinctly sluggish compressor with a fast attack can also help to set it apart from the rest.

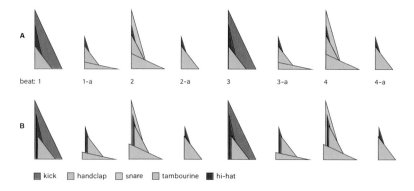

beat: 1   1-a   2   2-a   3   3-a   4   4-a

■ kick   ■ handclap   □ snare   ■ tambourine   ■ hi-hat

Figure 5.9 Parts don't need to line up exactly, as long as you give the listener a clear reference of where the center of gravity of the beat should be. 'Unquantizing'—moving tight loops slightly out of sync with each other— like in example B, can bring stacked parts to life, because it will be easier for your ear to distinguish them afterwards.

You can take this concept pretty far, as long as you end up with one well-defined, strong transient (usually the drums) taking the lead. In arrangements that sound a bit messy, I often solve the problem by choosing the most stable instrument to lead the rhythm, and making the others less important by attenuating their transients with compression, EQ, reverb and tape saturation. Or by shifting the messy part a bit backward in time, because your perception of timing is usually based one the first transient you hear, so that's the most important one.

Perspective Masking

Masking is an easily recognizable problem: you hear entire instruments disappear. Once you've detected it—and with a certain degree of skill— you can usually avoid it quite well through the use of balance, EQ and compression. It gets harder when the problem is less easy to recognize. Just imagine the concept you're trying to achieve with your mix. This is the creation of a world of sound that you connect to as a listener. In this world, you're mainly focused on the foreground, but all the surrounding elements together make it into a whole with a certain meaning. You maintain this illusion by having all the parts of the world relate to each other in certain proportions. Together, they form an image with perspective: their mutual proportions give the image meaning. A mix with a foregrounded bass drum and vocals lurking in the depths will have a very different meaning than a mix in which you make the vocals jump at you, while they're being supported by the same bass drum. The way in which you design this image is completely up to you, but the pitfall is

Assessing Compression

Our ears don't perceive transients as loud because of their short duration. This makes it difficult to hear how they relate to the longer sounds in a mix, which are easier to assess in terms of loudness. A good balance between these two is very important, because transients provide a lot of detail, depth and clarity to the mix. Not enough transients and those features will be lost; too many transients and the sound image loses calmness, strength and consistency. It's easy to reduce transients using compression or limiting, that's not the problem. But the trick is to decide how far you can go with this. Too much compression can make a mix dull, because the transients can't rise above the longer sounds anymore. If you then use EQ to compensate this, the longer sounds will become too bright and loud before the mix can sound clear again. When you assess transients, it can be helpful to vary the monitoring level. Listening at a high level for a short time can help you to feel if there's enough punch, without the sound image becoming harsh (to check if you haven't used too much compression), while listening at a low level can help you to hear if there's still enough detail (to check if you've used enough compression).

the same for every mix. If the foreground of the image moves around too much, the meaning of the rest of the world is constantly changing. As a result, you can't connect to the whole as a listener. You need a consistent foreground to guide you through the world (see Figure 5.10).

This kind of perspective masking—in which one inconsistent element distorts the perspective of the whole—lasts a lot longer than the 'normal' masking that has been extensively studied scientifically. This time, it's not about the inability of the ear to distinguish different sounds, but about the image of the music that you create in your head. If this image is disrupted, it can sometimes take minutes before you get used to it again. Especially long, sustained sounds with a lot of energy in the range your ears are most sensitive to (like vocals or loud cymbals) can disconnect you from the rest of the music.

Fortunately, compression is a perfect tool to prevent this. Because if there's anything a compressor does well, it's keeping things within limits. However, the compressor does have to 'realize' how the foreground is made up. For example, bass frequencies are less important there, but the range between 2 and 5 kHz all the more. Through sidechain filtering or

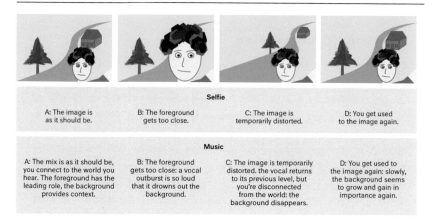

Figure 5.10 Perspective masking in music. If the foreground isn't consistently present, the meaning of the overall image distorts, just like it would in a photo.

even multiband compression, you can keep everything in this range within clearly defined limits, so you'll never be pushed away too far as a listener.

5.5 Dynamics and Loudness

Loudness is a powerful musical tool: if something is—or appears to be—loud, it will quickly have an impact. But whether a production will also be loud when it's played by the end user is usually not up to the engineer, since you can't control the listener's volume. What you can do is try to make the production seem loud by manipulating the perceptual loudness. The nice thing about perceptual loudness is that it works at low listening levels as well. This is why the average metal album also comes across as loud and aggressive at a whisper level.

A perceptually loud mix starts at the source, with instruments that are played with energy (and with the right timing), and whose energy is captured in the recording. From there, you can start emphasizing this energy in your mix, so the result will seem to burst from your speakers. The most important factor in this is the frequency balance. Your ears are most sensitive to (high) midrange frequencies, so this range is best suited for creating the impression of loudness. But it's a thin line between loud and harsh. If the music has a lot of long sounds in the high midrange—like cymbals, extreme distortion or high vocals—it will easily become too much. The trick is to find out exactly where the limit is, which sometimes means you have to attenuate a very specific resonance first, before you can emphasize the entire high midrange. Sometimes a bit of expansion can help: you add some more attack in the midrange, but you leave the longer sounds alone.

By partly blocking the sharpest peaks, compression can help to prevent harshness, while increasing the density of the sound. A small amount of distortion—for example tape or tube saturation—can also contribute to a sense of loudness. It adds some texture to the sound, a mechanism that's very similar to what happens when a musician plays or sings louder: guitar strings, speakers, drum heads and vocal cords generate more overtones when they have to work harder. They can't stretch any further, and as a result the peaks of the signal are rounded off. This is called acoustic distortion.

The paradox of arrangements that are meant to make a loud impression is that they're often filled with extra parts and bombastic reverb to make everything seem as grand and intense as possible, but that all these extras cost a lot of energy. So, in terms of absolute loudness, your mix of such an arrangement usually loses out to a small acoustic song. Perceptually it is louder, but in the acoustic song the vocal level can be much higher, because there's less getting in the way. That's why it's so pointless to try to make the average trance production equally loud (in absolute terms) as the average hip-hop production. It's an almost impossible challenge that can even prove counterproductive, because at some point the perceptual loudness will start to go down as you turn up the absolute loudness. The punch and power of the music will be lost, and these are the main things that contribute to the feeling of loudness.

If you feel that you can't make a mix loud enough without pushing your system into overdrive, you usually have an arrangement or (frequency) balance problem. An efficient use of the available energy is the key to creating a loud mix with impact. This means wasting nothing on pointless extra parts, on sounds that are partly out of phase or that take up too much stereo width, or on frequency ranges with no clear musical function. In the end, that's the sign of a well-defined mix: when every element is there for a reason.

Limiting

You might have noticed that in this chapter on dynamics—and even in this section on loudness—hardly a word is devoted to the subject of limiting. That's because of all the equipment in a studio, limiters are the least interesting when it comes to manipulating musical proportions. Limiters curb the peaks of a signal, and they're designed to do their job as inconspicuously as possible. Ideally, you can't perceive the difference before and after limiting. Therefore, it's usually not a very interesting process if you want to improve something audibly, and contrary to popular belief it's not a good tool for generating loudness either. Limiting

Limiting Basics

A limiter works like a compressor with a ratio of ∞:1 and a very short attack time. As a result, it will never let a peak exceed the chosen 'ceiling.' But this sounds easier than it is. A very short attack time means that, at the moment the limiter is engaged, it cuts the signal so fast that it can create sharp edges (see Figure 5.11). These sharp edges are distortion, and on some signals (like vocals or other long sounds) this can sound very annoying. The solution is to set the timing a bit slower, and have the limiter 'look into the future' with a buffer (look-ahead), so it will still be in time to block the peaks. But this results in audible 'pumping,' plus the transients will feel much weaker.

Look-ahead can work well for sounds that distort easily, but short peaks like drum hits sometimes sound better when you simply cut them off completely (clipping). This process has its limits,

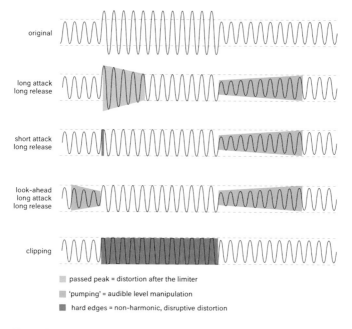

- ▨ passed peak = distortion after the limiter
- ▨ 'pumping' = audible level manipulation
- ▨ hard edges = non-harmonic, disruptive distortion

Figure 5.11 The pros and cons of different types of peak limiting. In practice, the original signal largely determines what the 'least bad' method is. For example, if there are a lot of short peaks, clipping sometimes sounds better than look-ahead limiting.

because beyond a certain point you'll clearly hear a crackling distortion that makes the entire mix tiresome to listen to. But if you use it in moderation, it can sound more transparent than limiting, because it doesn't introduce any timing effects (pumping). Some limiters combine both methods (a clipper for short peaks and a look-ahead limiter with a slow timing for long peaks) to reach the best compromise. But often the very best compromise is simply not to use so much limiting.

can increase the absolute loudness of a sound: by attenuating the peaks, you can make everything louder without exceeding the boundaries of the medium you use. But absolute loudness says nothing about the perceived musical loudness of a song (a drum-and-bass track will feel louder than the average ballad, for instance), only about the amount of energy that you fire at the listener.

In mixing, limiting is mainly interesting as a problem solver. For example, if you want to add a piano that's recorded so close to the hammers that every note starts with a loud click, limiting can help to better balance the sound. The same goes for tambourines, congas and other sharp-sounding percussion. The limiter can attenuate the transients, which can almost sound as if you're moving the microphone further back. Some limiters add a certain character to the sound when you crank them up: they can distort the sound or change the envelope in an interesting way. These kinds of side effects can be used as a creative tool, and in this role limiters can be musically useful. In section 14.4 you'll find more about loudness, limiting, and the considerations that come with it.

Chapter 6

Space
There's More Room than You Think

When I first started mixing, the results varied wildly. Especially when a band or producer wanted that bone-dry in-your-face sound, I'd often be less than pleased with how it turned out. I thought you created this sound by using no reverb at all and making every sound clear and compact. But comparing other mixes to my own, I found that they were in-your-face without sounding as dull and flat as mine. They sounded three-dimensional, but without the sound being very spacious or reverberant. I then discovered that even electronic music, which often has no acoustics, needs 'cement' to keep the music together in the right spatial relations. This allows things to not only be close or far away, but also big or small, for example. It gives the music perspective. And in-your-face music needs perspective too, because without it, it could never be in-your-face.

My next big discovery came not so long after that. I noticed that I could also make complex arrangements work by adding things, instead of taking them away. Until then, my first instinct had always been: 'There's too much information. I need to remove things or turn them down.' But this would usually result in a bland and boring mix. The next time I ran into this wall, I decided to try something completely different. I suspected that changing the character of some of the sounds would lead to a better blend. I messed around with reverb and delay, and to my surprise I could now easily pick out every single sound in a busy mix, even though I had only added an extra blur of reflected sound. For the first time, I realized that a mix has three dimensions, and that a deliberate use of these dimensions creates much more room to place instruments.

6.1 Hearing Space and Distance

Perspective

As described in the previous chapter on dynamics, music is communicated very differently through a recording medium than at a live performance. The impact of a recording is the result less of high sound levels, and more of contrasts in various sound properties. Depth is a great tool to create some of these contrasts, which can be especially useful to emphasize a musical transition. For instance, instruments can be far away during a verse, but then you suddenly 'zoom in' on them when the chorus begins. Differences in depth can also be used to support the hierarchy in an arrangement. Some elements will be more important than others, and placing them in the foreground is a good way to illustrate this. In a busy mix, you can't make the less important elements too soft, because they would be drowned out by the rest. But what you can do is place them further away. Then they will be almost as loud as the elements in the foreground, but still occupy a less prominent place.

Credibility

A natural perspective can make an electronic instrument or a sampled acoustic instrument more 'real.'It can be hard to achieve convincing results using such sample libraries, because besides expressive programming of the samples, a convincingly natural placement in the mix is also important. Especially if the sample manufacturer decided to record as cleanly as possible in an anechoic chamber, the user will have to reinvent the wheel. The same problem arises with DI-recorded electric (bass) guitars or synthesizers. These sounds refuse to blend in with the rest, until you place them in a credible perspective. Only then will they sound as if they ever really 'took place.' But in other cases it can be a creative choice to aim for a distinctly implausible perspective. For example, an almost head-splitting synth lead that has no connection with the rest of the music is a powerful statement. Mixes that sound 'too natural'—and therefore unremarkable—can be made much more interesting with an unnatural perspective.

Hearing Depth

So there are many reasons why it's useful to control the depth placement of sounds in your mix. In order to do this effectively, it helps to know how people estimate distances to sound sources. How do you actually hear if a sound is coming from far away? There are a couple of things that change when a sound source moves further away from you. First of all, the intensity of the overall sound decreases. Obviously, you can only hear this if the source is moving. If you don't know how loud it sounded at first, you can't say if it's becoming weaker. You might think this mechanism sounds exactly the same as a fade-out, but it's more complicated than that. The sound source creates reflections that bounce around the space, and the level of these reflections at the listener's position hardly changes when the source moves further away. What does change is the ratio between the direct sound and the reflections: the direct sound becomes weaker, while the reflections stay the same.

Therefore, if you want to determine your distance to the sound source, you need to distinguish between direct and reflected sound. Obviously, if a voice sounds very hollow, you know from memory that an up-close voice doesn't sound like that—and that this voice must be heavily affected by reflections. But luckily you don't even need to know what every source sounds like up close to make the distinction. You can find out all you need to know by combining four observations (see Figure 6.1):

1. Direct/reverb ratio: if the direct sound is louder than the reflections, you're close to the sound source. The inverse is also true: if the reflections are louder than the direct sound, you're far away from the source. The ratio is particularly noticeable for short, percussive source sounds, because you'll perceive the reverb's decay as clearly separate from the direct sound.

2. Directionality: the degree to which a sound is coming from one direction, or from all directions around you. You perceive this by interpreting the differences in time, intensity and timbre between your two ears. Direct sound originates from a single location in a space, while reflections come from all directions. Therefore, direction-ality is a good way to determine the direct/reverb ratio of longer sounds. Because these sounds overlap with their own reverb, it's harder to perceive the reverb separately.

3. Arrival time: direct sound always arrives at the listener before its reflections do. As a result, you'll easily recognize it as the source sound. But the time between the direct and reflected sound depends on the distance between the source and the listener. A big difference in arrival time indicates that the source is much closer to the listener than the walls of the space.

4. Timbre: the reflections arrive later than the direct sound, but they still interfere with it (the next chapter contains a detailed description of how this works). Certain frequency ranges add up while others are weakened, and this changes the timbre. Of course you don't know how the source originally sounded, but the sound of comb filtering due to strong reflections that arrive shortly after the source sound is very recognizable. Particularly in tonal sounds with a lot of overtones (strings, guitars, pianos, and so on) these kinds of colorations are easily noticeable. And then there's the effect that the direct sound and the reflections both become more muffled when they cover a longer distance to the listener. This is due to the absorption of the air and (partially) reflective surfaces. Especially if there are multiple sound sources in the same space, differences in timbre will give you a good idea of the distances between them.

The combination of these four observations works so well that you can even estimate distances when there are hardly any reflections. For instance outside, or in a room with very dry acoustics.

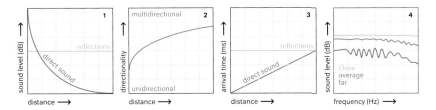

Figure 6.1   Four mechanisms to estimate the distance to a sound source: direct/ reverb ratio (1), directionality (2), arrival time (3) and timbre (4).

Size

Besides differences in distance between different sounds, there's another important tool for creating perspective. By playing with the size of the space in which certain sounds are located, you can create interesting contrasts. Just think of drums that seem to be played in a huge arena, while the vocals sing over them in a small room. The size of a particular space can be gathered from its time response (see Figure 6.2). If the reflections pile up right after the direct sound, but also decay quickly, the space is relatively small. Such a space doesn't have a discernible reverb tail. But if it takes a while for the reflections to build up enough density to form a thick reverberant field together, the space is relatively large. In such a space the decay is so long, that the reverb is clearly audible. The time it takes for the reverb to be attenuated by 60 dB is called the

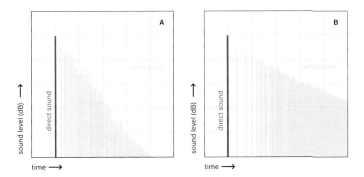

Figure 6.2   Time response of a small (A) and a large space (B).

reverberation time, or RT60. Some spaces are so big that you can perceive the first reflections as discrete echoes.

6.2 Simulating Space and Distance

An Interpretation of Reality

Up till now I've discussed how you perceive depth in natural situations. The mechanisms involved can be used to simulate depth in your mix. But not all of them can be translated directly, because listening to discrete sound sources in a three-dimensional space is not the same as hearing a recording of those sources coming from two speakers. This difference is crucial if you want to create depth convincingly with two speakers or a pair of headphones. To do so, you have to overcome three obstacles.

I. Masking Caused by Limited Directionality

Your depth perception is largely based on the distinction you make between direct and reflected sound. You can make this distinction partly because reflections come from all directions, and direct sound only from one. But if you record reflections in a natural three-dimensional situation and play them back in two-dimensional stereo, the area where they come from becomes severely limited (see Figure 6.3). As a result, the reflections are more densely packed and thus more likely to mask the direct sound,

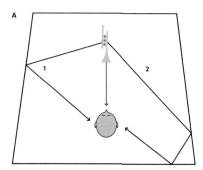
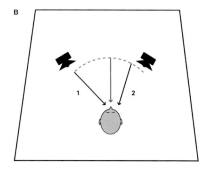

Figure 6.3 If a trumpet is played in the same space you're in, you can hear reflections (the black arrows) coming at you from all directions (A). This makes it relatively easy for you to distinguish the reflections from the direct sound of the trumpet (the red arrow). However, if you record this situation and play it back in stereo, the reflections will only be coming from the front, just like the direct sound of the trumpet (B). This will make it harder to distinguish the trumpet from the reflections, since everything is now more densely packed together.

making it harder to distinguish the two from each other. This is why you would normally place a stereo microphone setup closer to the source than where it sounds good in the room; or use a combination of close and distant microphones.

The challenge is to have reflections in your stereo mix that are effective in terms of depth perception, but without them masking the direct sound too much. In this respect, it would make your life a lot easier if you started mixing all your music in surround. But besides the fact that this would seriously restrict your audience, mixing in stereo—or mono—is a good learning experience. If you can master the trick on two speakers, the transition to five or seven speakers will only become easier.

2. Masking Caused by Production

Translating a natural situation to two speakers is not always what you're after. Sometimes you want to create a completely unrealistic—but no less impressive—sound image. In this case, the source sounds can take up so much space that there's hardly any room left for reflections. Adding realistic-sounding reverb to such a mix is counterproductive. The abundance of direct sound will only mask the densely packed reflections, so you'll never achieve the sense of depth you're after. It's better to use a couple of unrealistically wide-spaced reflections that are strong enough to provide some perspective amid all the noise, while leaving room for the direct sound.

3. Influence of the Existing Acoustics in the Recording and Listening Room

When you're recording you can't always foresee where you'll want to place a sound during mix time. But since you do record sounds in an acoustic space, there's always inherent placement in the recording—which will hopefully somewhat match the intended place in your mix. Material that's completely devoid of acoustics is a rare breed. Even electronic sounds often contain delay or reverb that already gives them a certain atmosphere and size. So in the majority of cases you'll be working with sounds that are already located in a space. On top of this, the listener will then play these sounds in a room that has its own acoustics. So if you want to manipulate the depth of the sounds in your mix, you should take the acoustics of the source sound and the listening room into account. Of course, both can vary widely depending on the situation, but it's good to realize two things. If a source sound was recorded in a smaller space or at a greater distance than where you want to position it in your mix, you're in for a challenge. And you shouldn't try to add a completely new acoustic space to the existing one (see Figure 6.4). As for the listening

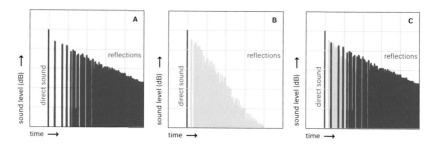

Figure 6.4 Adding the reflections of a large space (A) to a sound recorded in a small space (B) won't produce the same result (C).

room, you can be sure that more reflections will be added. These reflections are usually relatively short. You can take this into account by not using too many of these reflections in your mix. This will prevent your mix from sounding hollow when you play it outside the studio. Therefore, mixing in a studio that's too dry—or on headphones—can be deceptive, because you won't hear the added reflections of an average listening room. This also means that mixes that translate well to a wide range of audio systems often sound a bit overdefined on headphones.

So, with some adjustments, the same mechanisms that you use to perceive depth in a natural situation can be employed to give your mix perspective.

6.3 How Reverb Works

Long before music could be recorded or amplified, people were experimenting with using reverb as an effect. The composer would already imagine the church's reflections adding their part to the music while writing. Even in music that was performed in drier halls, you would occasionally find instrumentation that had to come from the balcony or behind the stage. The audience didn't have to get used to these kinds of sounds, as acoustic effects are very common in everyday life. You blindly trust your ears to interpret them, and you hardly notice them actively. This is why reverb is the most important effect in mixing. You can use it to create dramatic differences that will still sound credible to your audience. Unlike a whooshing flanger that immediately stands out as unnatural, an arena-sized reverb can easily be taken for granted. But it's still highly effective, because as soon as you turn it off, you notice how empty and meaningless the sound becomes. Of course you can take it so far that the reverb starts to stand out as an effect, but often this isn't necessary at all. Especially when reverb is barely audible in a mix, it can subconsciously play a major role. The great thing about working with reverb in this digital age is that you don't need an actual hall anymore to create a certain effect. All parameters that you need to go from broom closet to Taj Mahal are within reach. Because this can be pretty overwhelming at first, we'll first have a look at the various things you can influence.

Distance

Imagine you're standing in a space, and someone is walking toward you from the other end of the space, while singing continuously. What changes do you hear as the singer gets closer? Besides the fact that the direct voice gets louder while the reflections remain just as loud, there are also changes in time. The shorter the distance between you and the singer, the sooner the direct sound arrives at your ears. But the reflections of the space still take about just as long to reach you, regardless of where the singer is. So if a sound source is close to you, there's a relatively big difference in arrival time between the direct sound and its reflections. But if it's far away, the direct sound will arrive only a fraction earlier than the reflected sound (see Figure 6.5).

It's easy to simulate the difference in arrival time electronically with a short delay between the direct sound and the added reverb. This is called pre-delay, and it's a setting on almost all reverbs. Pre-delay is inversely proportional to the distance you're simulating: the longer the pre-delay, the closer the sound source seems to be. If you're employing reverb via an aux-send, for instance because you have one really nice hardware reverb unit that you want to use for multiple instruments, you can only use one pre-delay setting on the reverb. But maybe you want to run five different instruments through this reverb, but still position all of them at a different distance. In a DAW, you can easily solve this by creating a number of aux-sends, all with their own delay. The delay time should vary per aux and should be set to 100 percent wet and 0 percent feedback. Then you send the outputs of these 'pre-delay auxes' to a bus that sums them into your reverb. And if you also add some EQ to this construction to make the reverb sound darker at greater distances, you'll have a very flexible patch to simulate different distances within the same space (see Figure 6.6).

Size

Most artificial reverbs have a number of parameters that together determine the size of the space they simulate. One parameter they all have is reverberation time. This is the time it takes for the reverb to be

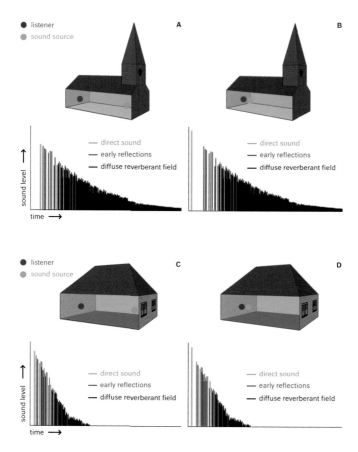

Figure 6.5 The way in which distance to the sound source and size of the space
affect the reverb.

A: The source is far away from the listener in a large space. The first reflections
arrive right after the direct sound, and are almost equally loud. As a result,
they add coloration to the direct sound.

B: The source is close to the listener in a large space. The first reflections arrive
significantly later than the direct sound, which is hardly colored as a result.

C: The source is far away in a small space. The first reflections arrive almost at
the same time as the direct sound, and are nearly as loud. Due to the short
distances, the reflections are close together, and as a result, the direct sound
is much more colored than in a large space.

D: The source is nearby in a small space. The first reflections arrive slightly later
than the direct sound, but still add coloration to it.

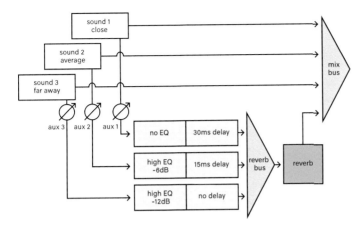

Figure 6.6 With this configuration, you can simulate three different distances with one reverb.

attenuated by 60 dB, after the direct sound has ended. To give you an idea: most living rooms have a decay time of less than half a second, while concert halls can sometimes reverberate for several seconds. This is because the walls in a living room are relatively close to each other, so the sound will hit a great number of surfaces in a short period of time. Besides reflecting the sound, these surfaces will absorb some of it, causing the sound to lose energy and decay relatively quickly. In a concert hall where the walls are far apart, it takes much longer for the sound to hit the same number of surfaces. As a result, it won't lose its energy as fast through absorption, and it will reverberate longer.

The pattern of the first (early) reflections is quite different in both of these spaces. Just take another look at Figure 6.5. The electronic parameter that corresponds to this pattern of early reflections is usually called 'size,' or it consists of a combination of density and build-up time. A greater density simulates more reflections per second, and therefore a smaller space, while a long build-up time suggests that it takes long for the reflections to blend into a thick soup, so the space is relatively large. Some reverbs that don't have a separate setting for size automatically vary the density and build-up time when you change the reverberation time. This is a bit restricting, because this way, you can't simulate a large space with a lot of absorption, and therefore a short reverberation time. There are also reverbs that let you select a different pattern of early reflections for spaces of different sizes. From small to large, the usual options are room, chamber, hall and church.

Generating Reverb

The oldest—and still very useful—way to add reverb to an existing recording is the echo chamber. This is simply a space with nice acoustics and a couple of speakers and microphones placed at a distance from each other. Then it's a matter of playing music through the speakers, maybe changing the microphone placement and adjusting the acoustics with partitions and curtains, and recording the resulting sound again. It works like a charm and sounds perfectly natural. The principle of using a real space to add reverb is also used in modern convolution reverbs. But instead of having to drive to the concert hall every time to play your mix and record it again, these reverbs use impulse responses of the building you want to simulate to manipulate the sound of your mix. An impulse response is the response of a space to a short signal that contains all frequencies equally (the impulse). You can see it as the fingerprint of the acoustics, which you later add to a digital recording. It's not possible to adjust the microphone placement or acoustics this way, but with a large collection of impulse responses, you usually have plenty of options.

Another way to generate reverb is simulating a real space with a model. The first attempts used electroacoustic mechanisms such as steel plates and springs that were excited into vibration and then recorded. These reverbs sounded nothing like a real space, but they did have their own character and strengths. Digital technology made it possible to generate increasingly complex reflection patterns. These so-called algorithmic reverbs had the advantage of much more flexible settings than echo chambers, spring reverbs and modern convolution reverbs as well.

Even though you can manipulate the time response of convolution reverbs quite well these days—for example, by only attenuating the early reflections, or delaying the late decay part—algorithmic reverbs (and spring and plate reverbs as well) still have an advantage when it comes to dynamic behavior. An impulse response is a static 'snapshot' of a space or a device. But if this device responds differently to soft and loud sounds, if it adds distortion or modulates parameters in time, this can't be captured with an impulse response. That's why both types of reverb (the real—or sampled—space and the simulation) have their place. Dense, natural-sounding reflections that won't even sound fake on percussion are better made with a real space or convolution reverb, but if you need a reverb that blends well with the mix and that should evoke a certain illusion, it's sometimes better to use a simulation.

Beyond Reality

The nice thing about artificial reverb is that you don't necessarily have to imitate a real space with it. For instance, analog plate reverb is still popular because it blends so well with acoustic recordings. This is because a plate reverb doesn't have early reflections like a real space, but it quickly generates a very dense reverb. Since most acoustic recordings already contain enough early reflections, plate reverb—plus a little bit of pre-delay—can complement these perfectly. But sometimes you want to take it a step further, for example by making the recorded space bigger than it is. If you just add a simulation of a real space on top of it, it will clog up your

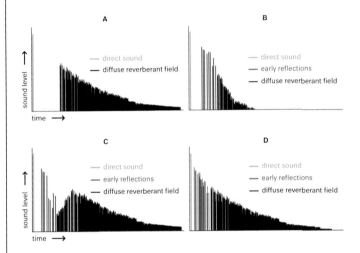

Figure 6.7 The possibilities of artificial reverb.

A: Plate reverb with pre-delay. The diffuse reverberation makes the direct sound seem grand, while the lack of first reflections combined with the pre-delay make it sound close.

B: The first reflections suggest a remote source in a large space, but the reverberation time is short, as it would be in a small space. The sound is far away in the mix, but you don't have the problem of a long tail making the sound too dense.

C: Reverb with less energy between 50 ms and 150 ms after the direct sound, which prevents the mix from getting clogged.

D: The pattern of early reflections implies a very small space, while the decay suggests the opposite. This reverb will sound very colored due to the high density of early reflections, and it's a good choice if you want it to attract attention.

mix. So it's important to be careful in your use of the space that you have.

This is what made Lexicon's classic digital reverbs so popular: they're so cleverly put together that they can create grand illusions with minimal means. David Griesinger, the founder of Lexicon, conducted a lot of research into the optimal distribution of reverb in time. He discovered that reverb in the time interval between 50 ms and 150 ms after the direct sound hardly contributes to a sense of size, while adding a muddy sound to a mix. With artificial reverb, it's a breeze to program it with less energy between 50 ms and 150 ms after the direct sound. This is one of the reasons why this type of reverb blends so easily with the rest of the mix. Figure 6.7 gives you an idea of the possibilities of artificial reverb.

Shape and Furnishing

An old factory building sounds very different than a concert hall, even though they might have the same size and reverberation time. This is because they are different in shape. The factory has a lot of straight surfaces that are exactly perpendicular or parallel to each other. Therefore, the reflection pattern created in this space will be very regular. Certain frequencies will resonate much longer than others, giving the reverb an almost tonal quality. This type of reverb will easily sound 'hollow.' A concert hall contains a lot more irregular surfaces, which scatter the sound in all directions. As a result, the patterns will be much less repetitive, and the space will sound more balanced. In other words: the reverb in the concert hall is more diffuse than in the factory (see Figure 6.8).

The electronic parameter that controls this effect is called diffusion, and it determines the amount of scattering. Old-fashioned digital reverbs didn't have enough processing power to generate a dense diffuse reverb, which is why these reverbs sound less 'smooth' than modern ones. But more diffusion is not always better: a part of the character of the room is lost when you make a reverb too diffuse. In the mix, it will be harder to hear what kind of space the sound is in. The acoustics become ambiguous and generic. That's why old digital reverbs are still popular: their character remains audible in the mix as a whole. This also goes for various analog spring reverbs, which are less successful at generating a diffuse reverb pattern than a decent plate reverb. A modern digital reverb with an adjustable amount of diffusion enables you to precisely control how smooth you want the reverb to sound. Or, in other words, how much attention you want it to attract.

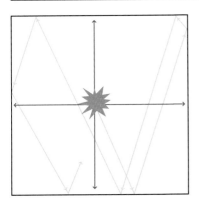 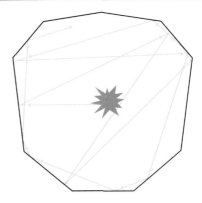

Figure 6.8 A space with a lot of parallel surfaces (left) produces regular reflection patterns, with some reflections going back and forth along the same lines (the red arrows). This will result in audible resonance. In a more irregularly shaped space (right) the reflection paths will be much more random. This space is more diffuse, and as a result it won't leave a very clear mark on the sound.

Besides the shape of the space, the use of materials has a great influence on the sound as well. A wooden floor won't sound the same as a stone floor. Curtains open or closed can also make a world of difference, and whether there's an audience in the room or not. All these things absorb sound to a different extent, and with different frequency characteristics. For example, an all-stone space will have more low and high frequencies in its reverb than an identically shaped wooden space. The effects of different materials can easily be simulated with electronic parameters for damping. The exact implementation of this varies per reverb. The most flexible solution is adjustable damping (or reverberation time) for each frequency range: low, mid and high. More damping produces a shorter reverberation time in the corresponding frequency range. This will sound significantly different than manipulating the reverb with an equalizer. For example, if you use a lot of high-frequency damping, the tail end of the reverb will be much more muffled than the onset. With an equalizer, the entire reverb would become equally muffled. Still, EQ is an often-used method for simulating damping as well. Besides parameters for damping, the most flexible reverbs have a separate equalizer for the entire reverb.

Cheating

Up till now, I've discussed how the parameters in artificial reverb correspond to real acoustic mechanisms. But as you have read in section 6.2, sometimes you need something other than real acoustics to make a recording convincing in stereo. A common problem is that you need so much reverb for the spatial illusion you want to create that it clogs up the mix. This is because the existing sound largely masks the reverb, so you will only hear the reverb when you make it too loud.

The main goal is a maximum illusion of spaciousness, while adding as little as possible. Therefore, what you do add should be masked as little as possible, which can be done by disconnecting the sound of the reverb from the existing sound. Modulating the length of the reflections is the ideal way to do this. The acoustic equivalent of this would be if the microphones in your echo chamber are constantly and randomly dangling back and forth. This creates minor pitch modulations in the reverb, causing it to be slightly out of tune with the direct sound. On solo passages of orchestra instruments this can result in an unnatural, chorus-like sound, but in a dense mix it works great.

You can also reduce masking by making the reverb come from a different direction than the direct sound. This is easy if it's a solo instrument that you just pan to the left and the reverb to the right, but in a complex combination of sounds it's not so simple. However, the reality is that in most mixes the most prominent elements are placed in the center. This means that there's more space left at the outer edges of the stereo image than in the middle. So basically you want to add more reverb to the edges than to the center, but how do you do this? Many stereo reverbs don't have a parameter for correlation, but luckily you can add one yourself.

Correlation is a measure for the relation of the left to the right signal. If left and right are exactly the same, the correlation is one. This is the effect of using a mono reverb and placing it at the center of the mix. If left and right don't have anything in common, the correlation is zero. An example of this is when you use two mono reverbs with different settings and pan them to the extreme left and right of the stereo image. This reverb will seem to come from all directions equally. If left and right are each other's opposites, the correlation is minus one. Left and right will have an opposite polarity then, and a reverb with this characteristic will sound so unnaturally wide that it seems to 'pull your ears apart.'

Most natural-sounding reverbs have a correlation that's a little on the positive side. If you use this reverb prominently in an open-sounding mix, it will create a convincing space between the left and right speakers. But in a dense mix, much of the reverb that seems to come from the center— with an identical presence in the left and right channels—will be masked. To avoid this, you'll have to distribute the energy of the reverb in such a way that the differences between left and right are emphasized a bit more. An easy way to do this is with a mid/side matrix, which is a feature of virtually every stereo widener.

Or you can send the reverb to an extra bus with reverse panning and polarity. You then add a little bit of this bus to the original, until the

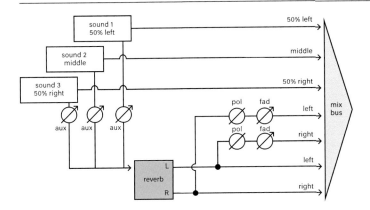

Figure 6.9 Besides panning the reverb wide, you can also send it to the opposite side of the mix, but at a low level and with reverse polarity. This way, you can place it beyond the speaker positions.

reverb is wide enough (see Figure 6.9). This way, you can even make the reverb come from beyond the speaker positions. But if you take this too far, your mix will collapse when you listen to it in mono. The acoustic equivalent of this method would be if you placed the microphones in your echo chamber further apart. In itself, this will make the stereo image of the space less stable (it creates a hole in the middle), but it will make it 'wrap around' the existing stereo image of the mix more easily.

6.4 Using Reverb

There is no light without darkness, no beauty without ugliness and no warmth without cold. Besides as content for self-help books, these clichés are also useful for mix engineers, because one of the challenges you'll face is the application and preservation of contrasts in your mix. For instance, if you use a reverb that makes almost everything sound bigger and better, it's easy to go overboard with it. Before you know it, everything in your mix is big and far away. And not even convincingly big and far away, because as the cliché goes, there's 'no far without close and no big without small.' So right off the bat, here's the most important piece of advice of this section: reverb can only have a real effect if there are distinctly dry and close elements as well. It's the difference between the reverberant and dry elements that determines the depth of the mix.

Noise

With some mixes it's hard to hear where their size and depth are actually coming from. In very densely arranged music like rock or trance, the reverb on one particular element often makes the other elements seem larger as well. Grand-sounding pad synths or dense, distorted guitars easily fill the space between the notes of the other elements in a track. In a way, they create a continuously present space that seems to contain all the elements. This mechanism can be very useful when the reverb on a particular element doesn't sound nice or clogs the mix, but you do want to build a space around that element. And if there's no continuously present musical sound that you can use for this purpose, you can even pull this off by adding noise. This might sound crazy, but it's a more natural scenario than you think. In virtually every space, whether it's a concert hall or a living room, there's background noise. Subconsciously, the soft hum of the ventilation, the wind in the roof, the patter of feet and the rustling of clothes gives you a lot of information about the kind of space you're in. When you softly walk into a quiet church, you'll immediately notice that you've entered a large acoustic space, even if there's no significant sound source present.

Recordings made in a quiet studio with microphones close to the sound sources are often unnaturally noise-free, as if they were recorded in a vacuum. The same goes for synthetic sources, which can sometimes sound eerily clean. In those cases, it can help to add ambient noise to enhance the realism. However, dry pink noise is not enough, as it contains no information about the size and sound of the space you want to simulate. But what you can do is filter a noise generator until you like the sound, and then run it through a reverb. Or even better: use a recording of the background noise in a concert hall.

Timing

To a large extent, the length of the reverb you can use to supplement the music is determined by the musical tempo and note density of the track you're working on. The quicker the notes follow each other, the shorter you have to make the reverb to prevent them from overlapping too much—otherwise they could form unintended chords. On top of this, the reverberation time also affects the groove: a short ambience reverb with a cleverly timed pre-delay can really give a rhythm a sense of drive. And a reverb that's just a bit 'too long' and decays right before the next beat can make a slow song feel even more sluggish and heavy. You can even think about the various parts of the groove, and how they might benefit from different reverberation times. Heavy accents get a long reverb, light ones a short reverb, and so on.

It gets even crazier when you start experimenting with reverb that begins not after but *before* the dry sound. Or for the real hobbyist: when you record the reverb of an instrument, cut it into pieces per chord, reverse all the pieces in time, crossfade them, and then place the whole thing a quarter or eighth note before the dry sound. The effect of these kinds of treatments is that you disconnect the reverb from the source, so the reverb will sound more like an individual element, plus it won't get in the way of the original sound so much. One of my favorite ways to achieve this kind of disconnection is by recording two takes of a vocal part, putting the first one in the mix and using the second one only to drive the reverb.

Stereo

The use of reverb has a big influence on the stereo image of your mix. Not enough reverb makes it feel disjointed, while too much will make it feel undefined. But it's not just a matter of more or less. The kind of reverb—and how you use it—can make or break the stereo image. In terms of width, artificial reverb comes in three flavors: mono, synthetic stereo and true stereo. Mono means that the reverb has a single input and output, synthetic stereo has one input and two outputs, and true stereo has two inputs and outputs. Since one way to recognize reverb is by the perceptual differences between your left and right ear, it would seem like the most logical choice to always use stereo reverb. Otherwise left and right would be exactly the same, making the reverb harder to pick out.

If you want to place a solo instrument in a natural space, this is definitely the right approach. But in the mix as a whole, you sometimes need a different effect. Mono reverb can be the solution if you want to use a lot of reverb because of the timbre and fullness that it gives to the sound, but the reverb itself shouldn't add so much to the total size of the mix. You can pan a mono reverb wherever you like. It will sound a bit like you're opening the door to a room (or a tunnel). In terms of stereo placement, you can attach it to the corresponding instrument, which will make the reverb stand out less, plus it won't take up so much space in the stereo image. It's as if you're placing the instrument in its own little cavity within the main space of the mix. But you can also pan it to the opposite side of the dry sound to create a dramatically wide stereo image, while the reverb as a whole takes up less space than a stereo reverb.

Synthetic stereo reverbs don't offer this kind of control. The sound that drives them always seems to be placed in the center, even when it's panned to the right in the mix. Which makes sense, given the fact that these reverbs only have one input. Sometimes it might look as if there are two, but under the hood, these two are first added together before they go into the reverb. In contrast, true stereo reverbs take the panning of the source sound into account. They generally produce a more defined stereo image than synthetic stereo reverbs, which makes them the best choice for adding reverb to existing stereo mixes, like overly dry classical recordings. On top of this, you can also do tricks with true stereo reverbs, like switching

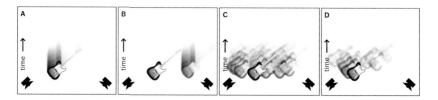

Figure 6.10 Different ways to use reverb in the stereo field:

A: Mono reverb coming from the same direction as the source sound.

B: Mono reverb panned to the opposite side of the source sound.

C: Synthetic stereo reverb coming from all sides, regardless of the placement of the source sound.

D: True stereo reverb that takes the position of the source sound into account.

the left and right channels of the reverb. This will mirror its stereo image compared to the dry sound, which can make your mix sound exceptionally wide.

Filtering

Reverb is the glue that keeps a mix together. Just like with a broken vase, you want the glue to fill the cracks without being visible itself. So the trick is to find the 'cracks' in the music to put some reverb in. For example, if you want to add a piercing vocal track to the mix, there are two approaches to choose from. You can use EQ or multiband compression to remove the sharp edges of the voice. This is guaranteed to work, but it will also cause the voice to sound unnatural and lack presence. Another option is to make the acoustics around the vocals more 'flattering.' If you put a reverb on the vocals that mainly emphasizes the warm-sounding frequencies, the direct voice will still sound natural while appearing less piercing and in-your-face. By carefully filtering the reverb, you can precisely determine how it will supplement the direct sound. The nice thing about this approach is that you don't need to manipulate the dry sound so much, keeping it more pure and powerful. It's like you're equalizing a sound in time.

An important choice to make is whether you should filter a sound before or after you run it through a reverb. In case of unpleasant peaks or imbalance in the dry sound, it's better to filter before the reverb. Many close-miked sounds are so clear and 'spiky' that it's more natural to the ear to make them a bit milder before you send them to the reverb. Otherwise these characteristics will only become more apparent because they last longer. For instance, reverb will add a long, hissing tail to overly

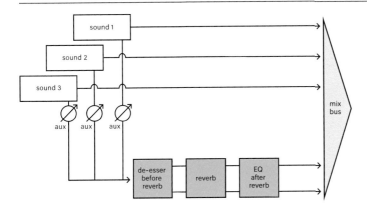

Figure 6.11 An example of filtering before and after reverb. The de-esser removes peaks before they become even more conspicuous due to the reverb's decay, and the EQ adjusts the character of the reverb to make it blend with the mix.

sibilant sounds, so you'd better tame the sibilance before hitting the reverb. But if the added reverb emphasizes a particular frequency range too much, it's often better to filter the reverb itself at its output. This is the equivalent of adding some acoustic treatment to a room.

To keep your options open, it's a good idea to always drive reverbs through aux-sends and don't use them as an insert, because this is the only way to manipulate the fully wet signal independently of the direct sound. On top of this, this method will allow you to keep the original signal in the main mix, while manipulating the dry signal that you send to the aux-send, for instance by cutting the high frequencies (see Figure 6.11).

Dynamics

In some cases, it can be difficult to successfully match a reverb with the source material. One problem you might run into is that the reverb is inaudible when the dry sound is soft, and too conspicuous when the source is loud. You can solve this by automating the send levels, but you can also put compression on the reverb. And there are more options at your disposal within the dynamics processing category. For example, you could use the sidechain input of the compressor to turn down the reverb while the dry sound is playing, and have it come back during the silences. This way, you'll have the decay of the reverb, while reducing the coloration of the dry sound. And if you want to have a huge reverb, but the decay is so long that it clashes with the tempo of the music, you can put a gate on the reverb. If you drive the sidechain input of the gate with

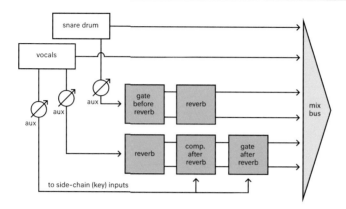

Figure 6.12 Examples of dynamic processing with the purpose of controlling reverb. Because the upper gate removes the leakage from the snare drum recording, only the hits are sent to the reverb. And the compressor (below) suppresses the reverb when the vocals start, but lets it come back in the open spaces. Finally, the lower gate determines how long the reverb's decay can be—regardless of room size.

the dry signal, you can use the hold and release time to precisely determine how long the reverb should decay during silences, and how suddenly it should be cut afterwards.

Gates can also work great if you want to use reverb on acoustic drums. Often, these recordings contain a considerable amount of spill, so it's not so easy to only put some reverb on the toms. Before you know it, there's reverb on the entire drum kit, turning it into an undefined mess. Running the dry sound through a gate to remove spill often sounds unnatural, but it can be a good compromise to only gate the signal that you send to the reverb. This will keep the reverb free from spill, while the dry sound stays intact (see Figure 6.12).

Extracting Reverb

So far I've only talked about adding reverb that wasn't present in the source material. However, a lot of acoustically recorded material (including acoustically recorded samples) already has reverb in it. Sometimes you have the luxury of separate room mics, which you can use to easily adjust the spaciousness (tip: experiment by shifting these microphones in time to make their effect bigger or smaller). But even if that's not the case, there are some tricks you can use to make the sound of the recording room more prominent. The method of choice for this is of course compression, which is an easy way to bring the loud direct sound and the weaker reverb closer together in terms of sound level.

Transient designers can also work wonders, plus there are various 'ambience extraction' processors you can use. These generally work by creating short, filtered delay lines that are stereo reversed. The delay follows the direct sound so quickly that the direct sound masks the delay. This is called the Haas effect. But the reverb that you couldn't hear before, can't mask the reverb in the delay. And voilà: you hear relatively more reverb now.

6.5 How Delay Works

You might have noticed that a lot of mixing techniques discussed in the previous chapters are designed to counteract masking. When you start putting several sounds together, you're bound to lose details that you'd rather keep. So far, we've employed equalizers, compressors and panners to prevent this. But on top of that, we've also used reverb to add more perspective to the mix. And although this works well, it also has an unwanted side effect: it creates more masking.

On an Old Fiddle . . .

It's nice if you can create perspective while wasting as little space as possible. You want to maintain the illusion that reverb creates, but rid the reverb of everything that's unnecessary. In this respect there's a lot to learn from the time when you couldn't simply use twenty reverb plugins in your mix. In the fifties and sixties, not every studio could afford a plate reverb or special echo chamber. But they all had a tape recorder, so this piece of equipment was used a lot to create a 'suggestion' of reverb. With a tape loop, it's easy to generate echoes: the longer the loop or the lower the tape speed, the longer the echo. And when you softly send the output of the tape recorder back to the input, the resulting feedback creates a pattern of repeating echoes.

This might sound more like a game of squash than a beautifully reverberant cathedral, but in the mix as a whole it's surprisingly effective. So effective in fact that tape echo is still used a lot today. Sometimes as a prominent effect in alternative music, but more often it's used in such a way that you don't notice it as a listener. And then this corny Elvis effect suddenly becomes a building block that mixes in all genres can benefit from. It's like taking a reverb that consists of tens of thousands of reflections, but then you only hear a few of them, which is enough to give you an idea of the size of the space. When used subtly, it can even add some convincing depth to classical recordings, without the density that reverb would cause.

As so often the case with good inventions, chance plays a major role in their success. The reason why tape echo sounds so good is the fact that the system has shortcomings. The echo is not an exact copy of the original, but it has a slightly different frequency spectrum (due to distortion), pitch (due to minor tape speed fluctuations), and reduced dynamics (due to tape compression). This is not exactly the same as what happens to a sound bouncing back and forth in a real space, but it's pretty similar. Due to absorption and diffusion, the timbre of acoustic echoes also changes compared to the original sound. The advantage of echoes that sound different from the original is that they won't be masked so much. This means that you can turn them down in the mix, yet still achieve the same effect. At a lower level, they will interfere less with the original sound, which will sound more natural and powerful as a result. Try adding a simple echo with feedback to a vocal track by using a clean digital delay without filtering or modulation, and then compare it to a tape echo. You'll notice that the tape echo can be softer in the mix and still create the same sense of spaciousness, which means it won't get in the way of the vocals so much.

Tape is not necessarily the holy grail when it comes to echo. Any process that changes the timbre of an echo can be useful. Think of EQ, compression and distortion, but reversing or pitch shifting the delayed signal can work as well. If you use feedback, the change of sound will become more intense with every new 'echo round.' For example, you can make an echo sound darker with each repetition by attenuating the high frequencies in the feedback loop (see Figure 6.13).

Another popular delay is based on analog memory chips. These bucket-brigade delays can hold an electric charge for a short time before releasing it again. If you place enough of these buckets in a sequence, together they can hold an electric charge for quite some time. You could see this as a from of analog sampling. The sound of these chips is far from perfect: the high frequencies are often lost completely, and with each repetition, the signal gets darker and grainier. But here, these imperfections help the sound as well: even the distortion that results from driving the chips a bit too hard can really help to give the echoes their own sound, which in turn will make them sit in the mix more easily.

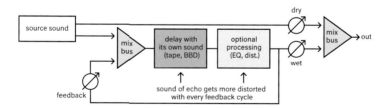

Figure 6.13 Block diagram of a delay with feedback. Due to shortcomings of the delay system, or thanks to additions by the designer, every generation of echoes becomes increasingly distorted.

Multitap

All the delays discussed so far have been single delays that use feedback to create a repeating pattern of echoes. This works very well for almost all applications, but if you want to create more adventurous rhythms than repetition patterns made up of equal steps, you'll quickly run into limitations. In these cases, multitap delays can provide the solution. Multitap means that you tap a delay system (or delay line) not only at the very end, but also at three quarters of its length, for instance. This will result in a repeating pattern of two quarter notes of silence, an echo on the third quarter note and another one on the fourth. With tape echo, you can achieve this effect by positioning a number of read heads at a distance from one another (see Figure 6.14). Digitally it's even easier to create the craziest patterns. Before you know it, a single handclap becomes a complex rhythm that stretches across four measures, with each echo having its own filtering and panning.

A modern take on the multitap delay is spectral delay, which divides the signal into a large number of frequency bands (let's say 1024). You can then run each band through its own delay with feedback, before you put everything back together again. This way, you can create otherworldly sounds that have hardly anything to do with the source sound. You won't need this in every mix, but it offers great opportunities for sound design

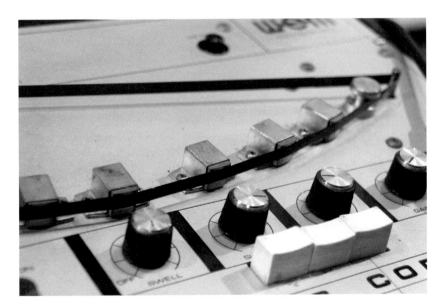

Figure 6.14 Multitap delay the analog way: this Watkins Copicat has multiple read heads that pick up the signal from the tape at different moments.

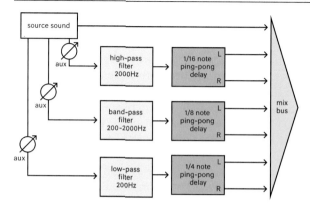

Figure 6.15 An example of a matrix that divides the sound into three frequency ranges, each with its own delay. This construction can also be used for all kinds of other effects that you only want to use in a specific frequency range.

(you can find more about these spectral manipulations in section 9.6). If you're looking for a similar—but less extreme—effect, you can create a setup like this yourself with some filtering and delay plugins. It's easy to make a division into three frequency bands, each with its own delay or another effect (see Figure 6.15).

6.6 Using Delay

Timing is Everything

You can use delay in two ways: to give sounds depth or to create an audibly rhythmic effect. The difference between these two approaches is mainly determined by the delay time you set, and this can be quite tricky. If the delay time is too short, the resulting comb filtering will add a lot of coloration to the dry sound, and if it's too long, the echo will be audible in itself, which could cause it to clash with the timing of the music. If you just want to add depth, you can take the Haas window as a starting point. This is the time interval between approximately 5 ms and 20 to 40 ms after the direct sound. Delayed copies arriving within this interval aren't perceived as individual sound sources by your brain, but as part of the direct sound. The maximum delay time you can get away with before you start perceiving a discrete echo depends on many factors. The loudness, timbre and panning of the echo play a role, as well as the tempo and note density of the music. But the type of instrument also makes a difference: with percussion, 20 ms of delay can already be too much, while strings can sometimes take as much as 100 ms.

With rhythmic effects, it's a very different story: anything goes. The way you use delay in these situations is mainly a matter of taste, but it can be a nice experiment to let go of the quantized (tightly synchronized with the tempo of the music) rhythmic grid for a change. Setting the delay time manually opens up a world of possibilities. For example, you can make the entire rhythm feel slower and heavier by setting the delay to a quarter note plus an extra 20 ms. Or vice versa: an agitated song will sound even more hurried if the echoes seem to 'rush' a bit as well. In some cases it can even work to completely disregard the tempo and rhythm. If the instruments are already continuously emphasizing the rhythm in a song, a delay that perfectly locks with the rhythm will only have a marginal effect. On the other hand, if the echoes end up randomly between the notes, they will stand out much more. Sometimes this can be annoying, because it distracts from the rhythm too much. In those cases, it can help to make the sound of the echoes more diffuse. You can take the sharp edges off by attenuating the high frequencies, or by adding extra effects like chorus or reverb. Once you get the hang of it, you'll get away with huge amounts of delay without the listener consciously noticing it. If you were to press the stop button halfway through the chorus of an average pop song when it's being mixed, you'd be surprised at how much is still resounding. You didn't notice the delay in the mix as a whole, but when you turn it off, the mix will suddenly become small and flat.

Panning

Where you place the echo compared to the original (dry) sound is crucial to its effect. Short echoes that come from the same position add coloration to the direct sound. The original sound and the echo will be barely distinguishable from each other, blending into a new sound source that seems further away than before. So echo coming from the same direction as the original sound strongly affects your sense of depth and distance. But if you place the echo opposite of the original, the effect is very different. Since the original sound is much less affected, it won't move away from the listener as much. In this case, the delay has the perceived effect of moving the original sound outward, further than you could achieve with panning alone. In a way, you're applying the principle of time-of-arrival panning from Chapter 3. As a result, it sounds as if the original sound is placed all the way to the front and side of a wide space. The delay time determines the width of this space. That is, until you make it too long and the illusion collapses. At that point, you'll suddenly hear a separate echo instead of a space. Figure 6.16 shows various ways to pan echo, and the impact on the original sound.

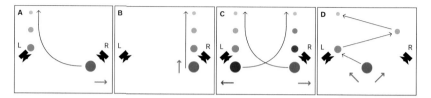

Figure 6.16 The effect of panning a delayed signal (smaller dots) in relation to the original sound (large dots). The colored arrows indicate the perceived effect on the original sound. If you pan the echo to the opposite side of the original sound, the latter will seem to move further outward (A). If you place both on the same side, the original sound will move backward (B). The effect of A can also be used to make stereo sources even wider and more spacious (C). And if you want to make a mono source wider, a so-called ping-pong delay can work well (D). These are two mono delays that are panned wide and that give each other feedback, causing the echoes to bounce back and forth across the stereo image.

Using mono delays isn't as bad as you might think, because panning them gives you a lot of control over their effect on the original sound. Many delay plugins are stereo by default, and their presets pull out all the stops to sound as wide as possible. Sometimes this works, but it often has a less powerful effect in your mix than a 'focused' mono delay. Even the effect in Figure 6.16D of echoes ping-ponging back and forth between the left and right channels can be achieved with two separate mono delays. The advantage of this method is that you have full control over the stereo image. You can make one of the echoes come from the far left and place the other one softly in the center. This way, you can easily make the effect less wide, or focus it more on one side of the stereo image.

Ducking

Sometimes it's hard to set the delay in such a way that the echoes only fill the 'cracks' in the mix and not more. For example, if the part you want to put delay on has a rhythm that keeps changing (with alternating short and long notes), the echo will sometimes blend in with the mix and sometimes clash with the original part. This can be solved by using automation to adjust the amount (or even the length) of the delay to the part. This way, you'll know for sure that it works well everywhere. And once you get the knack of it, you'll be able to manipulate your echoes like a real Jamaican dub mixer, emphasizing musical accents or even creating new ones. But if you don't have the patience for this, there's another trick, because it's also possible for sounds to influence their own delay level. By putting a compressor on the delay and driving its sidechain with the dry sound, this sound can 'duck' the delay. The compressor suppresses the echoes when the part is playing, and turns them up again during the open moments.

The only parameter you'll find on a lot of delay effects that I haven't discussed yet is modulation. But this changes the effect so much that it's separately addressed in section 9.3.

Chapter 7

Time and Phase
It's All About the Sum of the Parts

A mix is a combination of sound sources. As soon as you put two or more different sources together, they start to influence each other. You could say that mixing is nothing more than using this influence to your advantage. For example, you can make one source seem more important than the other. As discussed in the previous chapters, this can be done by manipulating the balance between parts, as well as the placement, timbre and dynamics of the sources. However, there is another determining factor for the influence that two sound sources have on each other: the interplay between time and phase.

7.1 Interaction

Sound is a pattern of successive air pressure variations that move through the air, away from their source, at a speed of about 340 meters per second. Every time a sound hits a hard surface it bounces back, so at a single position in a space you can hear the sum of the original sound and thousands of reflections. The time it takes for the reflections to reach your ear determines their influence on the original sound. Something similar happens when you record a drum kit with a microphone close to the drums, and combine this track with a room microphone that's placed at a distance. The sound of the drums will first reach the close microphone and then the room microphone, and the time difference between these microphones determines how the sum of the two turns out. This time difference causes some frequencies of the drums to arrive at the microphones 'in phase,' and others out of phase. But what does this mean?

Phase

Phase is a measure (expressed in degrees or radians) that indicates the position of a repetitive wave in its cycle (period). For example, after a quarter of a second, a sine wave with a frequency of 1 Hz (which completes its entire cycle once per second) is also at a quarter of its period.

Another way to express this is in degrees: an entire period is 360 degrees, so a quarter period is 90 degrees. Of a 1 Hz sine wave, you can say that it takes 0.25 seconds to shift 90 degrees in phase. In other words: phase is a measure to express a frequency-dependent time interval, because 90 degrees for the 1 Hz sine equals 0.25 seconds, while the same 90 degrees for a 2 Hz sine is only 0.125 seconds (the frequency is doubled, so the period is halved). Therefore, a phase shift can only mean something if you mention the frequency(ies) involved (see Figure 7.1).

Frequency

Due to this frequency dependence, phase isn't very suitable as a measure of time, but it's very useful if you want to convey the mutual influence of two waves with the same frequency. For example, if these waves are 180 degrees apart (which can also be the case if one is at 360 degrees of its cycle and the other at 540 degrees), they will cancel each other out when you add them to your mix at an equal amplitude. When this happens, the waves are 'out of phase,' and their frequency will be completely inaudible. However, if one is at 360 and the other at 720 degrees, they are in phase, so they will amplify each other (by 6 dB).

This is why phase is a good concept if you want to describe how two waves of the same frequency interact with each other when you mix them, because it says something about the extent to which they amplify

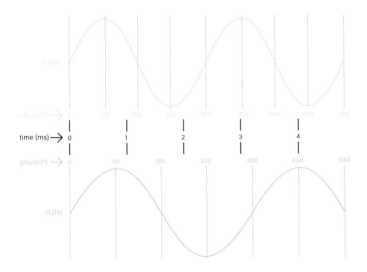

Figure 7.1 The phase number in degrees indicates the position of a wave in its repetitive period: an entire period is **360** degrees. For each frequency, this period has its own length, which is why a phase number corresponds to a different time interval for each frequency.

or attenuate each other. In practice, only pure sine tones consist of a single frequency, as most instruments produce an entire frequency spectrum, and each of the frequency components in this spectrum can be seen as a sine wave with a certain amplitude and phase (see Figure 7.2).

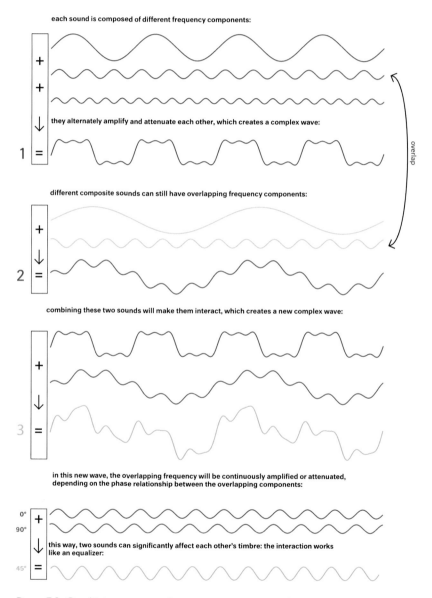

Figure 7.2 Combining two sounds creates an interaction between the overlapping frequency components, which has the effect of an equalizer.

Correlation

When you combine two separately recorded sound sources in your mix, their mutual influence is random. A simultaneously playing kick drum and bass guitar both consist of a number of frequency components that overlap here and there, in time and frequency. When they overlap, the phase relationship between the overlapping components determines the outcome. If they are out of phase (a 180-degree difference) they can attenuate each other completely, and if they are in phase (a 0- or 360-degree difference) they can amplify each other by 6 dB. But usually the phase difference is somewhere in between. The overlap of frequency components—and therefore their interaction—is completely random in this example: every different tone the bass guitar plays changes the overlapping frequencies and the interaction with the kick drum. But even if the bass guitar always plays the same tone, the phase relationship between the overlapping frequency components is still random: due to constant, subtle changes in the timing of the kick and bass, the addition or attenuation is different for every note they play together. There's no fixed connection between the bass guitar and kick drum (unless you sample and loop them, of course). Another way to express the random relationship between two signals is by saying that they aren't correlated.

The opposite happens when two signals correspond to a certain extent: then they are correlated. If they are completely identical, they amplify each other by 6 dB when you combine them. In this case, the value of the correlation is one. But if the two signals are each other's exact opposites (mirrored copies), the correlation is minus one, so they completely attenuate each other. In practice, you won't come across these extremes very often, but signals that strongly resemble each other without being exact copies or opposites are anything but rare. When sound reflects in a space, you'll hear more copies of the same sound, all subtly different in timbre and arrival time.

These (early) reflections are correlated, and combining them can produce extreme filtering effects, depending on the difference in arrival time (see the box on comb filters). These effects are so extreme because they aren't random, and they don't continuously change with the music, as uncorrelated sources do. It's always the same frequency components that add up and attenuate. The result sounds like a static filter, and it's easy to recognize as an unwanted coloration. The same problem can occur when you use multiple microphones at different positions in a space to record the same source (like a drum kit), or when you combine a DI signal with a miked amplifier. Therefore, this chapter specifically deals with controlling the interaction between correlated sources, because it's only there that these types of filter effects can occur.

Comb Filters

A comb filter is exactly what you imagine it to be: a series of successive amplifications and attenuations in the frequency spectrum that looks like a hair comb. These patterns are caused by the inter-action between a sound source and a slightly delayed copy of the same sound source. Such a situation can occur when a sound is reflected off a nearby surface, or when you use two microphones placed at a short distance from each other. Two sounds always interact when they are combined, according to the principle in Figure 7.2. But when the sounds are copies of each other, the result of this interaction is no longer a random amplification and attenuation of frequencies. Instead, a very recognizable, unwanted coloration occurs (see Figure 7.3).

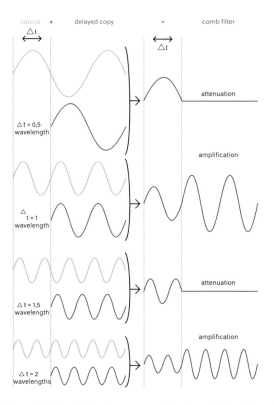

Figure 7.3 Comb filtering as a result of combining the original signal (green) with a delayed copy (blue). Depending on their wavelength, certain frequency components will either attenuate or amplify each other (red).

Just try to create this effect in your DAW by copying an audio track and adding it to the original with a millisecond of delay. It's good to get familiar with the resulting sound, because this will make it easier for you to recognize and solve acoustical and electrical problems. Make sure to also experiment with the level difference between the original track and the delayed copy. You'll notice that if the difference is bigger than 10 dB, you can't really hear the comb filter anymore. So this is also the level from where spill can become a problem: if a hi-hat mic captures a snare drum at a level that's more than 10 dB lower than the snare's own microphone, you probably won't experience any problems when combining the two.

Comb Filters and Direction: Your Brain versus Your Mixer

If the original sound ends up in your left ear and a delayed copy in your right, the resulting comb filter between those two sounds will be less of a problem than if they come from the same direction. This is because your brain doesn't add the sounds together the same way a mixer does: instead of a single, extremely filtered frequency spectrum, you hear the average of the two frequency spectra that arrive at both ears. This is why stereo recordings with two spaced microphones (time-of-arrival stereo) still don't cause a disruptive comb filter (and definitely not on headphones). The comb filter will only occur when you listen to these recordings in mono. In those cases, it won't be your brain but your mixer that combines the time differences.

Time

The degree to which a sound source and a delayed copy are correlated is determined by the duration of the delay measured against the sound's variability. With a percussion instrument—which produces a rapidly changing frequency spectrum—it takes a relatively short time until it's no longer correlated with later arriving echoes of the same sound. By the time the reflections arrive, the spectrum of the source sound has changed so much that the frequency components overlap in a more random (uncorrelated) way. With a choir that sings very long, constant notes this is very different. For a long time, the source and its reflection have a nearly identical frequency spectrum, so they will correlate. In general, the shorter the delay between the source and its copy, the stronger they will correlate and the greater the negative influence (in the form of a comb

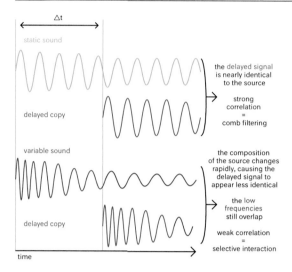

Figure 7.4 The temporal nature of a sound partly determines the amount of interference it will suffer from reflections or additional microphones. With drums, comb filters are much more apparent in the high frequencies on long-resounding cymbals than on short snare drum hits. With many instruments, the low frequencies aren't very variable because they're usually tuned to a certain tone or resonance. These low frequencies will resound a bit longer, and are therefore more sensitive to the influence of reflections.

Spill

Spill, the leaked sound of one instrument that's picked up by the microphone of another instrument, can be a pretty nice addition to the direct sound, as long as it arrives quite a bit later (otherwise comb filtering occurs) and at a lower level than the direct sound. These two aspects can be tackled at the same time by placing the instruments relatively far apart and positioning the microphones relatively close to the instruments. Because sound spreads spherically, every doubling of the distance to the source reduces the level by 6 dB. Indoors, this rule doesn't really apply due to reflections, but any added distance helps to reduce spill. You can also take the instruments' radiation characteristic into account, as a drum kit radiates sound quite evenly to all directions, while a trumpet's sound is much more directional. If you aim the trumpet away from the drum kit, there will be significantly less spill ending up in the drum microphones than if you play in the direction of the drums. Beware of hard, reflecting surfaces that act like a mirror: if the trumpet is aimed away from the drums, but straight at a wall, there will still be a lot of spill in the drum microphones. Setting up the instruments at an angle to the wall can give better results in this case (see Figure 7.5).

Microphone Selection

When placed correctly, microphones that pick up sound directionally (cardioid, hypercardioid or figure-eight microphones) can help to reduce spill from other instruments. Just as important as the direction a microphone is facing is where you point its insensitive area (the 'null'), and how this area sounds. Cardioid microphones often don't sound very nice on the sides and back (off axis), which means that they do suppress leakage from other instruments, but the remaining spill can sound very bad. Sometimes you'll achieve better results with omnidirectional microphones. This approach will make the spill relatively loud, but at least it sounds good. Only if the microphones can't be placed far enough apart for practical reasons, or if the acoustic balance between the instruments isn't perfect (unfortunately, both are common scenarios), might it still be necessary to use directional microphones.

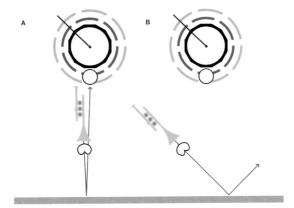

Figure 7.5 Choosing a smart recording setup will help to attenuate and delay spill as much as possible before it reaches a second microphone.

filter) will be. That's why, when placing microphones, you try to prevent the same source from reaching different microphones with very short time differences.

Checking Polarity

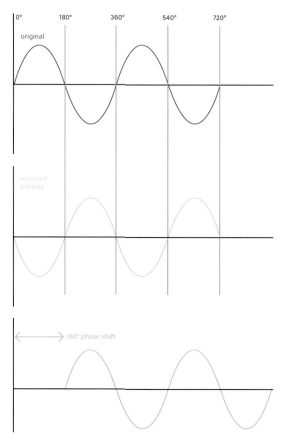

Figure 7.6 The difference between reversing polarity and adding delay: in the case of this sine wave, the added delay in the bottom figure creates a phase shift of 180 degrees. This causes the wave to be in antiphase with the original, just like in the middle figure, where its polarity is reversed. However, with a sine of any other frequency, the same delay will cause a different phase shift, so this other frequency won't be completely in antiphase with the original. This is the principle that causes comb filters, and the polarity switch on your mixing console can only correct some of it, as it will only shift the frequencies of the resulting comb filter. This can be slightly more favorable or unfavorable for the overall sound, but it's much better to prevent comb filtering altogether.

When you combine two correlated sources in your mix, you should first check their polarity. Polarity is an absolute value (plus or minus) that says something about the 'direction' of a signal. Imagine a stereo system where the plus and minus wires of the right speaker cable are reversed. If you send a positive mono input voltage to this system, the left speaker will move forward, while the right one simultaneously moves backward. In technical terms, this means that the polarity of the right channel is reversed, or that all frequencies are 180 degrees out of phase (see Figure 7.6).

Testing polarity is the easiest thing to do: simply press the polarity switch of one of the channels involved, and check if the sound improves or deteriorates. An improvement is usually audible as an increase in low frequencies, and the opposite is true for a deterioration.

If you can't get a nice combined sound with the polarity switches alone, it's possible that the various tracks aren't aligned in time. This can be due to two microphones that are placed far apart, or to latency caused by a digital processor. In this case, a short delay, or simply shifting one of the tracks forward or backward (in many DAWs, this is called the 'nudge' feature) can be the solution. Zooming in on the waveforms in your DAW can be a useful visual aid in deciding how much nudging is necessary. An ideal match of the source material in terms of polarity and time will help a lot to make your mix more powerful. Still, it's smart to wait a bit before you start aligning all your drum microphones, because the story gets a bit more complex.

7.2 Combining Microphones Hierarchically

The most realistic-sounding recordings use only a stereo microphone that picks up all aspects of the sound (directness, spaciousness, definition, timbre, dynamics) in the right proportions. Admittedly, it does improve the spaciousness if you use two microphones placed some distance apart, but that's about as far as we will go. This is the classic recording philosophy: the fewer microphones you need to make a balanced recording, the fewer problems you'll run into when you combine them, and the more realistic the sound image will be. Recording everything with just two microphones is a nice ideal, but as with many ideals, it's not always possible to realize this. You'll often need additional microphones to improve the balance, definition and illusion of space. Or maybe realism is not what you're going for, and you want to have full control over the sound image by using separate microphones for each instrument. For example, heavy metal drums only work when you use microphones to zoom in on certain aspects of the drum kit, and then exaggerate these in the mix. But it's not always easy to make all those microphones work together well. Why is that?

When you use multiple microphones to record the same source, these microphones have a certain mutual relationship, call it a marriage. There are two basic principles that define this marital bond: the law of the loudest and the first-come-first-served principle. Although this might seem barbaric, it's completely natural. Just imagine two speakers producing the same sound simultaneously and with equal loudness, but one is a meter away from you and the other six meters. The sound from the nearest speaker will be the first to reach your ears, and it will also be louder. Therefore, it will dominate the sound image. This domination means that our brain gives this sound priority. When the exact same sound arrives at the listener twice in rapid succession, the first—and loudest—sound will be interpreted as the source, and the later arriving weaker copy as a reflection, as originating from the same source. This principle of copies that appear to merge into a single sound source is called the Haas effect, also known as 'the law of the first wavefront.'

First Come . . .

The direction of the first-arriving sound determines the perceived location of the sound source. Your brain assumes that the first sound comes directly from the source, as opposed to a reflection that takes longer to arrive. The reflection greatly contributes to the perception of distance, but you don't perceive it as an individual sound source. In the example of one close and one distant speaker, the first speaker can be one meter away to the right and the other six meters away to the left, but you still hear both speakers as a single source coming from the right. This illusion is only broken if the difference in arrival time between both sources is big enough to cause an audible echo.

The time frame in which two sources merge into one is called the Haas window, which ranges from about 10 to 50 milliseconds. The exact value depends on the type of source sound. On drums, you'll hear the echo sooner than on strings, so 50 milliseconds will be too long for percussive sounds. For a better visualization, you can use the speed of sound (340 m/s) to convert the Haas window to distance: 10 milliseconds is equal to 3.4 meters and 50 milliseconds equals 17 meters.

Translating this knowledge to a two-microphone recording of a single sound source means that the difference in distance between both microphones and the source can usually be about ten meters before you start hearing an echo. However, this rule only applies if both microphones are equally loud in the mix. If the source is weaker in the distant microphone than in the close microphone, the delay can be a bit longer before the listener starts perceiving discrete echoes. For instance, blending room microphones placed at 10 meters from a drum kit with close microphones will only cause problems if the room mics are extremely loud in the mix. You can reduce the risk of audible echoes even further if you make the room mics less percussive (with a limiter, for example) or more muffled, because high-frequency peaks (transients) play a large role in the identification of individual sound sources.

In a recording situation for classical music, it works the other way around: the main microphones—which are placed at a distance—are generally the loudest in the mix, even though the sound will reach them later than the supporting spot microphones. These spots may be less loud in the mix, but because they're the first to pick up the sound, the Haas effect works to their advantage. Since classical recording technicians usually prefer the timbre and image of the main microphones, they want our brain to prioritize these. In a way, the brain should interpret the main microphones as direct sound, and the spot microphones as an addition. Therefore, the sound of the main microphones should actually be the first to arrive at the listener. Besides being weaker, the sound of the spots will then also arrive later, which will give them a more supporting role in determining the timbre and placement of the instruments. This effect can easily be achieved by delaying the spot microphones. But beware, because choosing the ideal amount of delay still requires some thought.

There's No Such Thing as Equality

Of course, you can try to perfectly align all the microphones. This seems like the logical thing to do, as the main microphones are the loudest, and though the spot microphones arrive at the same time, they will be less loud in the mix. As a result, the main microphones will mainly determine the sound of the mix in terms of timbre and stereo image. Sadly, the reality is a lot more complicated. Since the two main microphones are usually placed some distance apart, there will be time differences between the left and right recordings. The sound of an instrument playing at the right side of the stage will first reach the right microphone and then the left. If you want to boost this instrument with its own close microphone, the only way to really align this spot mic with the main microphones is by aligning it separately with the left and right main microphones. If you don't do this, the spot mic's signal will only add up in phase with one of the two main microphones, and the sound image will be blurred.

The correct approach would be to perfectly align the spot microphone with the right main microphone, route it to its own stereo bus, pan it slightly to the right, and finally give the left half of this stereo bus a delay that corresponds to the difference in distance between that specific instrument and each of the main microphones. However, following this approach means you would have to do this separately for all the spot microphones. This is a lot of work, plus it doesn't work well in practice. The reason for this is that spot microphones never just pick up one instrument, but also the instruments and reflections around it. You can only synchronize a single instrument with the main microphones, and everything else the spot microphone picks up will be slightly out of sync (see Figure 7.7).

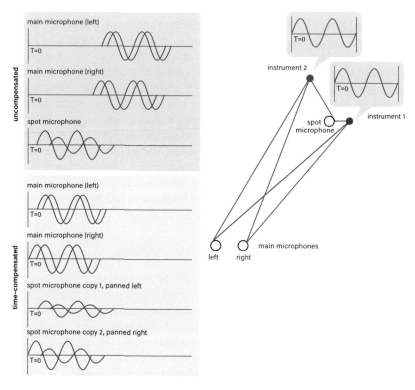

Figure 7.7 Aligning a spot microphone with two main microphones has its limitations. The top right image shows a microphone setup for the recording of two instruments. In addition to the main microphones, instrument 1 has its own spot microphone. However, instrument 2 is also picked up by this spot microphone. In the gray box (top left), you can see the signals of the individual microphones, and the blue box shows the same signals after you've aligned them. Instrument 1 (the red sine wave) can be synchronized with both the left and the right main microphone, by copying it and aligning it separately with the left and right channels. As a result, the time difference between the two main microphones remains intact, and therefore the stereo image as well. The problem is in the spill of instrument 2 in the spot microphone: it's relatively soft (the blue sine wave), yet still sufficiently present to cause a partial attenuation of instrument 2 in the main microphones. This is due to the fact that instrument 2 is further away from the spot microphone than instrument 1. Therefore, you can never align both instruments with the main microphones.

This slight deviation is exactly what's so annoying. Preferably, microphones that pick up the same sound should either align perfectly in terms of timing (and thus add up in phase), or differ significantly (and thus have random phase interactions). This is because everything in between causes comb filtering when both signals are combined. Therefore, if you're recording a single, stationary sound source with two microphones (like a guitar amplifier), it definitely makes sense to align these microphones. But as soon as there are more sound sources playing at the same time and even more microphones—or when the musician is moving slightly—you're usually better off not aligning the microphones.

Hierarchy Works

Now you understand why it's also futile to align different drum microphones, because every solution creates a new problem. It's better to look for a workable time difference between the microphones, preferably already during recording. This means that the spot microphones should be delayed so much compared to the main ones that there's no audible comb filtering. But at the same time they shouldn't be so late that they become audible as discrete echoes or start cluttering up the mix. Of course, this also works the other way around. For example, when recording a drum kit in a small space, it's better to delay the room mics enough to reduce their filtering effect on the direct sound, but not so much that you start hearing an echo.

As mentioned before, the amount of delay that you can add between microphones before you run into problems depends on the source material. This is why I can't give you any hard numbers, only a starting point. If you start by aligning all microphones as much as possible (compensating for the differences in distance due to microphone placement), you'll often end up placing the secondary microphone(s) between 10 and 50 milliseconds after the prioritized microphone(s). By using the nudge function in your DAW, you'll quickly find the right amount of delay.

This principle also works the other way around. For example, if the main microphones lack definition in terms of placement, you can move the spot microphones ahead of them, instead of delaying them behind the main ones. This way, they can give little clues about the position of the instruments in the stereo image. These clues are mainly provided by the first onsets of instruments, the transients (see Figure 7.8). If these peaks only just barely cut through the mix ahead of the main mics, it will be enough to provide definition in most cases. Now the spot microphones are the first to be heard, so they don't have to be as loud as when they're aligned with the main microphones. On top of this, they emphasize the transients because they're placed so close to the source.

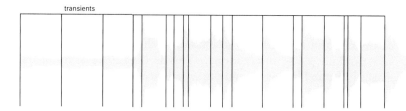

Figure 7.8 Transients are short peaks that usually mark the beginning of a note.

7.3 Combining Microphones Non-hierarchically

So far we've mainly looked at the relationship between main microphones and spot microphones. These are by definition unequal, as one is meant to complement the other. But this all changes when you start recording instruments with multiple microphones placed relatively close to each other, as the signals of these microphones will be very similar. Just think of recording a snare drum and hi-hat with two microphones, or using two or three microphones to record a singer who's playing an acoustic guitar at the same time. Of course you'll try to avoid recording the same source with more than one microphone, but because the sources (the snare and hi-hat, or the voice and guitar) are placed so close together, there will always be a lot of spill, and all microphones will essentially record both sources.

In these cases, the microphone signals correlate, and need to fulfill equal roles in the mix, which is why it's important that they add up nicely. It's not an option to delay the guitar microphone by 20 ms to reduce any comb filtering. If you do this anyway, it's mostly the spill of the guitar in the vocal microphone that will define its sound in the mix, as this spill will arrive at your ears before the direct sound does. To avoid this, you should try to get both microphones as perfectly aligned as possible. So before you start fixing it in the mix, it's better to first look for an acoustic solution.

Prevention

When recording vocals and acoustic guitar simultaneously, the ideal setup would be if the guitar and vocal microphones recorded the guitar and vocals at the exact same moment. This would produce a perfect addition of the source sounds and their spillage, without comb filtering. Mathematically, there's only one way to do this: by placing the guitar and vocal microphones at the exact same position, so there won't be any differences in distance. In the most extreme case, you only have one mono microphone to record both the vocals and guitar (it should be clear by

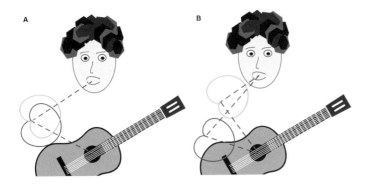

Figure 7.9 Recording a singer/guitarist with coincident cardioid microphones
eliminates differences in distance, so the spill of the guitar in the vocal
microphone is perfectly aligned with the guitar microphone, and vice
versa (A). However, if you use spaced microphones, differences in
distance (B) do occur, which often result in comb filtering.

now why this is not a bad idea at all). However, this means you won't
be able to change the balance in the mix. But you can also think of a
vertical X–Y placement with two cardioid microphones, one aimed up
toward the voice and the other down toward the guitar. Or better still:
the same setup, but with two figure-eight mics (Blumlein), which are even
more directional than cardioids. The problem is that it's not easy to find
a single position where both the guitar and the vocals sound good, as a
lot depends on the qualities of the musician, the instrument, the acoustics,
and the intensity of the music. Usually, you'll succeed at balancing the
volume ratio between the vocals and guitar, but if you want to change
the spatial perspective or timbre, your options are very limited. That's
why the problem is often solved with spaced microphones, one close to
the vocals and one or two on the guitar. However, this does mean more
work for the mixer.

Cure

Even with microphones that have a high directivity, there will still be
vocal spill in the guitar microphone, and vice versa. If you place both
microphones from Figure 7.9B in the center of the mix and make them
equally loud, you'll hear a delayed copy of the guitar and vocals (the
spill) interfering with the direct microphone signal. If the singer/guitarist
doesn't move, this interference will result in two static filters, but usually
you'll hear a subtly modulating comb filter effect caused by movement.

This will be particularly audible in the vocals, as most listeners have a strong reference of how vocals normally sound.

For this reason, you should first try to eliminate the comb filter effect on the vocals as much as possible, and then hope that the guitar will still sound acceptable. You can do this by shifting the guitar microphone slightly forward in time until the vocal spill is aligned with the direct vocal track. Sometimes you're lucky, but often it's not possible to really make this work due to slight changes in the singer's position. In those cases, the best solution is to equalize the guitar microphone in such a way that the most disturbing vocal spill is attenuated a bit in the guitar microphone. If you hear a whistling resonance around 8 kHz when you add the guitar to the vocals, you can try to attenuate this frequency in the guitar microphone. This way, the guitar mic will mostly add warmth to the guitar sound, while interacting less with the vocals.

Cheating

If alignment and EQ alone can't do the trick, there's another escape route: panning. By nature, stereo recordings made with an A–B setup are full of minor time differences between the left and right channels, and yet they often still sound great. It's not until you press the mono button on your monitor controller that things start to sound thin and small, because then it's not your brain but your monitor controller that's responsible for combining the two channels (see the box on comb filtering in section 7.1). Following this principle, you can use panning to prevent the two conflicting signals from being added electronically, and leave it to your brain to perform this addition. This doesn't mean that you should completely pan the vocals to the left and the guitar to the right, but every little bit helps. If you pan the vocals 20 percent to the left and the guitar 40 percent to the right, there will already be less interference compared to panning both microphones in the middle, as there's less overlap now. There's a reason why I'm only mentioning this escape route now, because it's really worth your while to first make the two signals add together as best they can.

This is because non-aligned signals can be pretty difficult to pan. Let's say you have two microphones, and you want to place one in the center and pan the other one to the right. If the right microphone has a slight delay compared to the center microphone, the combined sound will seem to come from the left! This is a common pitfall when combining two microphones that recorded the same guitar amplifier. In some cases, the time difference between the two microphones will perfectly complement the stereo placement, eliminating the need for alignment. Just listen and choose the time difference between the two microphones that best matches the position where you want to pan the sound, and you're done.

This technique works best if you pan the microphones all the way to the left and right, while occasionally performing a mono check to see if the sound isn't becoming thin.

7.4 Phase Manipulation

Sometimes you can nudge all you want, but the two signals simply refuse to line up. For example, when a bass guitar is recorded both directly and through an amplifier. This won't cause any problems with spill (like in the example of the singer-songwriter), as it's simply two isolated versions of the same sound that don't move compared to each other in time. Then why is it still so difficult to align them? The problem lies in the distortion and filtering added by the amplifier, which changes the sound of the bass guitar. Filter effects in particular make it difficult to get the amplifier signal perfectly in phase with the direct signal, as filters don't have a straight phase response. This causes a phase shift, which can be seen as a frequency-dependent delay, changing the internal relationship between the various frequency components of the bass guitar. If you listen to the amplifier signal on solo, you'll barely notice these phase distortions. They only gain meaning when combined with the undistorted direct signal, because this creates unpredictable phase interactions (see Figure 7.10).

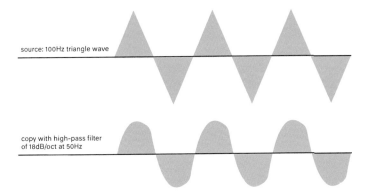

source: 100Hz triangle wave

copy with high-pass filter
of 18dB/oct at 50Hz

Figure 7.10 After you run it through a 50 Hz high-pass filter, a 100 Hz triangle wave suddenly looks very different. The various frequency components that make up the triangle wave have shifted away from each other: this is called phase distortion. Therefore, time alignment of these two waves is only possible to a limited extent. This particular case won't be much of a problem, but the more filters you add, the harder it gets to achieve good alignment, especially if these filters have steep slopes (in Chapter 17 you can read more on the drawbacks of filters).

The Last Resort

What can you do about this? In any case, you should start by synchron-izing the tracks as best you can. Usually you can use a transient as a reference point for this, and in the case of Figure 7.10 you would simply align the zero points of the waveforms. Now you're ready to start experimenting with an all-pass filter. These filters don't change the frequency relations, but only the phase relations in a signal. Ideally, by varying the all-pass filter, you'll find the exact inverse of the phase distortion caused by the recording chain. This will make the bottom wave in Figure 7.10 look like a perfect triangle wave again. Or, if you take the filter you found and inversely apply it to the top wave, you can make it look exactly like the bottom one.

Sadly this is impossible in practice: the phase distortions are too numerous to be compensated precisely, so you wouldn't know where to start looking. But in many cases, what you can do is improve to some degree how the two tracks add up. This is a very intuitive process. As said before, phase distortion can be seen as a frequency-dependent delay. With a variable all-pass filter you can delay the frequencies around the filter's cutoff frequency with the turn of a single knob. So it's simply a matter of running one of the two recorded channels through the all-pass filter, turning the knob and listening carefully if the sound of the combined signals improves or worsens (see Figure 7.11). Keep in mind that any EQ that you apply to one of the channels afterwards will cause phase distortion again, so you might need to adjust the all-pass filter.

An all-pass filter can be seen as a frequency-dependent delay. In select cases, it can help to align frequency components better. Figure 7.11 shows a (very schematic) representation of the two main frequency components picked up by an overhead and a snare drum microphone when both the snare and the tom of the drum kit are being played. If you don't do anything with these two signals, the snare drum will be partly attenuated when you mix them together (A). Reversing the polarity of the snare drum mic will improve the situation for the snare, but now the tom is attenuated (B). Delaying the snare drum microphone gives a similar result (C), but after combining this delay with an all-pass filter that only delays the main frequency of the tom (D), both frequency components are in phase. In reality, this scenario is much more complex, as both the snare drum and tom contain many more frequency components. Plus you shouldn't forget that the microphones also pick up the spill created by other elements of the drum kit. Therefore, you should ask yourself if using an all-pass filter doesn't just create a bunch of new problems while it's only solving one. The only way to assess this is by listening. If you don't detect a problem to begin with, there's no need to start shifting tracks in time or turning all-pass filter knobs.

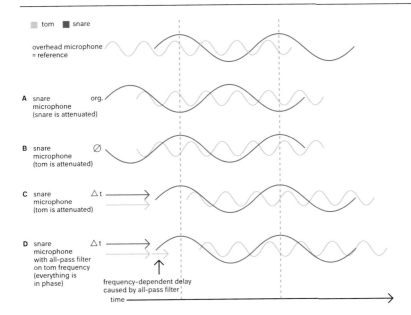

Figure 7.11 A possible use of all-pass filtering.

Chapter 8

Identity
A Unique Sound

Identity can make the difference between music and muzak. Is it completely forgettable, or is there something that appeals to the imagination, that surprises you, makes you sad or cracks you up? Of course, those qualities need to be in the music to begin with. They should be embedded in the notes and the performance. But if they aren't amplified by the recording and the mix, chances are they will go unnoticed by the listener. That's why a mixer should have a keen ear for artistic identity, and develop ways to support it—or even shape it—with sound.

8.1 Innovation, Imitation and Inspiration

Sculpting audio identity starts with the musical context. What mood is the music trying to set? What is it about? And what are the key elements in creating that meaning? Which elements make the piece distinct and which make it powerful? The easiest way to give a piece of music a 'face' is to make the unique elements relatively loud in the mix. Such a division into major and minor elements makes for a more pronounced mix, though you can also take this too far. Finding the right balance between making listeners feel comfortable and drawing their attention is important. If you're too avant-garde, you might scare away new listeners, which wouldn't do the music a very good service. But if you're not unique, listeners won't find your music worth their time, and the same thing happens. What works well is referring to other productions without literally imitating them. Often, you'll already do this automatically, as your own frame of reference is the starting point for everything you do. So the main trick is to not imitate or repeat too much, because this will result in elevator music. Therefore, ways to inspire new choices are very welcome in the studio.

Chance?

Besides hitting the bong, there are many other things you can do to spark new ideas. Introducing an element of chance can be very refreshing. This doesn't mean you should follow John Cage and start making all your decisions by the toss of a die–rather not actually. Instead, think of small things like taking the manipulations you applied to one channel, and using them for a completely different channel as an unexpected starting point. Or try to create a mix with only drums, bass and vocals, and make it sound like a finished product. If you then add the other parts on top, you'll automatically treat them very differently than if you had added them from the beginning. Quickly setting up rough mixes—and if it turns out they don't work, immediately throwing them away and starting

Figure 8.1 Innovation, inspiration or imitation? If your design is too innovative, no one will know you're selling peanut butter. But if you simply copy your competitor's formula, you won't make a lasting impression. With peanut butter—not a product consumers so strongly identify with that they start wearing brand T-shirts—this is not a problem, but with music it definitely is. The trick is to create something unique that your audience will still understand.

afresh—is a good way to find the best mix approach. Working together with different people also helps to come up with new methods, as long as you avoid ending up with a weak-sounding compromise that's meant to please everyone.

A nice way to force creative thinking is by limiting the options. If you only have a guitar to make your music with, at some point you'll start exploring the limits of this instrument. Imitating a cello with your guitar will produce a more distinctive sound than a cello sample. In these digital times, it's often a deliberate choice to adopt constraints, as there are hardly any left technology-wise. It takes quite a bit of self-control to look for one unique-sounding synth patch, instead of stacking three that are mainly impressive, but lack personality as a whole. The same principle goes for the use of effects: stacking three reverbs is often a lot less powerful than giving one element exactly the right reverb while keeping the rest a bit drier. Using equipment with restrictions or forcing yourself to do everything with only a couple of plugins can help you to stay close to the essence of the music. On top of this, it stimulates you to find ways around the constraints, which in turn can produce unique sounds. Too many options can result in a jumble of half-choices and contrasting ideas: ten cool ideas can sound horrible when put together. Many mixes only have a couple of sounds with a very strong identity, which is sometimes created or enhanced with effects equipment. The other sounds have a more supporting role, both in sound level and in the absence of a pronounced sonic character. Identity attracts attention, and usually you don't want to spread this attention to all parts evenly.

Back in the Day . . .

I'm not saying that you should willfully fence yourself in by buying an old four-track tape recorder, just so you can be creative through limitations. It's often said that the Beatles would have never sounded the way they did if they had had 24 tracks at their disposal. And while this is true without a doubt, it's also important to see how technological development stimulates creativity. At the time, the Beatles saw the arrival of multi-track technology as a huge increase in options, inspiring them to try new production ideas. Maybe we have too many options now, but new technology can still produce new ideas. It's mostly about what you don't do with all the options you have. It's true that the limitations the Beatles were forced to work with shaped their music. But you can also say that they by no means utilized every option available. Production-wise, every choice was made in service of the songs, and that's no different now.

In music production, musicality and vision are as important now as they were before. You can add, repair and fake anything you want, but the music will only be good if you know when it's okay to do this, and when you had better not. I think the main difference between now and the analog age is that it's no longer the norm to work together and take the time to get into a creative flow, and—just as important—connect a deadline to that period. When you're working on your own at your computer, it's harder to stay inspired. You'll get distracted more easily, as you'll often have several projects or other activities running at the same time. Plus it's hard to say when something is done, and when you should stop adding more effects.

8.2 Shaping Identity

It's not easy to find something that's special or different. Putting a weird effect on the lead vocals can definitely make your mix sound unique, but not necessarily in a good way. There should be a connection between the sound and the content. But what does this mean? When is a feedback-heavy flanger related to the content, and when is it far-fetched kitsch? On the one hand, it depends on the genre, the era and the current fashion, or in other words: it helps if you have good taste. But on the other hand, it's also about guts and persuasion: if you dare to give the flanger a defining role that also feels good musically, you can get away with the craziest things. As long as it contributes to the mood of the music—whether it's meant to make you dance or evoke strong feelings of sadness—it can work. You'll only go too far when the sound starts to attract more attention than the music. Sometimes it can be nice to create a short accent this way, but if a mix is so full of effects that you lose track of the melody or rhythm, it's not a good mix.

In this regard, it's interesting to look at what gives sounds their identity in everyday life. Besides the character of the sound source itself, there are resonances, filter effects, reverberations and echoes caused by the acoustics around it (see Figure 8.2). These kinds of mechanisms can be simulated very well by using effects equipment. The good thing about this is that you can lay it on pretty thick before your audience starts tuning out. People are so used to acoustic coloration that they hardly perceive it as an effect. Reverb, filtering and delay are much more accepted than (quickly) modulating effects like tremolo, vibrato, phasing, chorus or flanging. The nice thing about reverb and delay is that it doesn't need to sound very realistic to be effective. You can use it to make distinct statements that are still convincing. It wasn't until after the eighties that people started wondering if all that fake reverb wasn't way too kitschy after all. But at the time, the audience never complained about it: the music just sounded contemporary to them. They never consciously noticed all the white noise and gated cathedral reverb that was used to beef up those eighties snare drums.

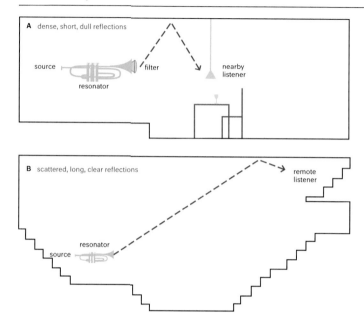

Figure 8.2 The same trumpet can have a totally different identity if it's played with a mute in a small club (**A**) or at a distance in a reverberant concert hall (**B**). Through the use of electronic filters and reverb, you can create similar differences in sound that feel very natural.

Musical function and sound can be seen as two separate things. As long as the musical function (bass line, rhythm, melody or harmony, for example) stays the same, it's okay to restyle the sound that performs this function. In essence, arrangement is the distribution of musical functions among instruments in order to give a composition a unique color. In the studio you can do this as well, by looking for sounds that might not be the obvious choice for a certain role, but that do have the right sonic character for it. Anything you can bang or tap can be used as a percussion instrument. One of the most beautiful bass drums I've ever discovered turned out to be a plastic garbage can in a reverberant space. And even if you don't go out and explore the world with a field recorder, you can use effects to give your source material a certain musical function. A pitch shifter can turn a singer into a double bass or bass drum. By making a delay with feedback oscillate, you can create a tone out of any given source sound. With distortion or compression, you can turn a short peak into a ringing, sustain-heavy sound, and so on. As long as you keep the musical function in mind, anything goes. Especially if you find a balance between your experiments and more familiar sounds, you can offer listeners something they'll understand right away, while still appealing to

Selling Effects with Acoustics

Reverb is the ideal tool to make strange sounds seem natural. Suppose you stuck a cutout of Donald Duck among the figures in Rembrandt's *Night Watch* if you had the chance. It would be clear that it doesn't belong in the picture. But if you make Donald match his surroundings in terms of light, shadow, hue and perspective, he'll seem to become part of the whole—albeit in a stylistically dubious way. This principle is true for all kinds of strange, electronically generated sounds that need to find a place in your mix. When placed in a convincing space, they will sound a lot more natural, as if they've always been there.

It's important to be clear when you do this: if an effect is a statement, it shouldn't be too soft. Apparently you have an artistic reason to stick a cartoon figure onto the *Night Watch*, so be brave and flaunt it. However, it can also be meant as a hidden twist, and the power of the effect should be in its near-subconscious subtlety. In that case, Donald should be hardly noticeable. Spectators will see a familiar picture, but feel there's something beneath the surface. They will have to look closer to notice what's going on. In both cases it's clear what the audience should or shouldn't immediately pay attention to. This is also important in music: if an effect doesn't have to be a pronounced statement, it's usually a bad thing if you can easily point it out. In such a case, it should be supporting the music rather than leading it.

their imagination. For example, a natural-sounding vocal line can help to 'sell' a very experimental accompaniment.

Dare

The hardest thing about trying to make something unique is staying mindful of the world around you—without being afraid to be different. The digital distribution of music has made it so easy to compare different productions, that many makers spend a lot of time emulating the 'competition' in terms of sonic impact, loudness, frequency balance, and so on. As a result, the products they create are more and more alike. It takes guts to deviate and present the world with a production in which the lead vocals are barely audible, and which has a pretty dull overall sound. Still, among the music that moves me the most, there are songs that actually sound like that (and that might just be worldwide hits as well). That's what I try to hold on to when I make decisions that make the sound less impressive, but that hopefully enhance the intensity of music.

Especially at the end of a production process, it gets harder and harder to stay true to your original vision, as your perception of musical identity will have waned considerably by then. You've been working on the individual components of the song too much, losing track of what it's like to hear it on the radio for the first time. During this final stage, the overall loudness and frequency balance can still be objectively compared to other productions, but these things are much less important than the musical impact. This is why you won't see a producer like Rick Rubin in the studio a lot: he simply doesn't want to lose his musical judgment before the project is done. At select moments, he gives instructions to his production team, who take care of the details. This way he can keep an eye on the big picture.

If you don't have a Rick looking over your shoulder, it's good to remember that it's not a smart strategy to try to please everyone with your mix, as this means not choosing a clear direction, but sticking to the middle of the road. This approach is fine if you're making background music for commercials or call centers, but not if you hope to claim a unique place in the world with your music. Therefore, it's better to pick a direction that will impress some people (hopefully a lot), while making others cringe.

Chapter 9

Effects
Craziness with a Purpose

Acting crazy for the heck of it tends to amuse only one person: yourself. A mix with special effects flying all over the place is like a bad joke that only makes the joker laugh.

This chapter explores the means that you can use to make many bad sound jokes. But they can also help the music a lot, if you use them sensibly. Effects can attract attention, but they can also work more subtly, like the clues you subconsciously pick up in a Sherlock Holmes book. You can direct the listener's attention with them, give extra meaning to certain things, and use them to build tension or make suggestions.

9.1 Distortion: Simulating Intensity

Using distortion is like coloring outside the lines, which might be exactly why it sounds so interesting. When you push a device beyond its ability, it can result in the most unpredictable, raw sounds. This evokes a kind of primal response in the listener, who perceives the sound as loud, dominant and aggressive, even if it's soft in absolute terms. But also when it's used as a subtle effect, in barely perceptible amounts, distortion can be a very useful tool to give a mix more cohesion and warmth.

9.1.1 How it Works

Overtones

An acoustic instrument produces sound by generating and amplifying a vibration. Plucking a stretched string, blowing on a reed and hitting a membrane are ways to excite such a vibration. The acoustics of the instrument then determine which vibration is amplified through resonance. For example, the soundboard and connected soundbox of an acoustic guitar are designed to evenly amplify as many notes as possible, just like the acoustics of a good studio. The tube of a clarinet, on the other hand, was developed to produce one specific resonance (depending on which tone

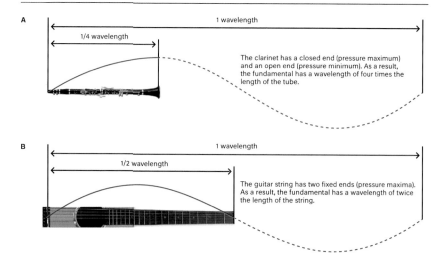

A   1 wavelength

1/4 wavelength

The clarinet has a closed end (pressure maximum) and an open end (pressure minimum). As a result, the fundamental has a wavelength of four times the length of the tube.

B   1 wavelength

1/2 wavelength

The guitar string has two fixed ends (pressure maxima). As a result, the fundamental has a wavelength of twice the length of the string.

Figure 9.1 Fundamentals.

hole you open). A quarter of the wavelength of this resonance's frequency fits exactly between the reed and the first opened hole (see Figure 9.1A). In a guitar, it's not the soundbox that determines the pitch, but you who does it by adjusting the string length with your finger. This creates a vibration with a wavelength that's exactly twice as long as the distance between the bridge and your finger (see Figure 9.1B).

Harmonic

The vibrations in Figure 9.1 are the lowest frequencies that can resonate in the instrument; these are called fundamentals. But there are more frequencies with wavelengths that precisely fit the instrument's system, which can therefore resonate as well. These frequencies, which are integer multiples of the fundamental (for instance two, three or four times the fundamental frequency), are called harmonic overtones, and they form beautifully harmonic intervals with the fundamental (see Figure 9.2A). Many sound sources have a similar series of harmonic overtones, albeit in varying compositions. You're so used to the sound of this phenomenon, that if you hear a combination of harmonic overtones, you automatically add the corresponding fundamental in your head. The standard waveforms on a synthesizer (sine, triangle, sawtooth, square wave) can all be seen as a fundamental supplemented with a specific series of harmonic overtones (see Figure 9.2B).

Non-harmonic

Not all resonances in an instrument relate to the fundamental in neat intervals. Sometimes the shape of the soundbox or the vibration mechanism produces a series of overtones that seem unrelated to the fundamental. In church bells or drums you do hear a tone, but it's much harder to identify than the same note on a piano. This is because these instruments contain a dense pattern of non-harmonic resonances above their fundamental. These kinds of overtones make for a less 'smooth' and sometimes even unpleasant sound. But higher harmonic overtones can also sound dissonant, for instance when they relate to the fundamental in intervals containing minor seconds.

Identity

Whether they're harmonic or not, overtones give a sound its identity. If you have old strings on your guitar, the higher overtones attenuate

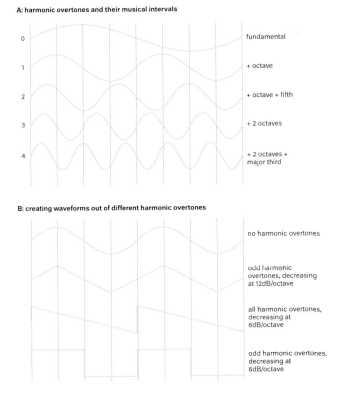

A: harmonic overtones and their musical intervals

0 — fundamental
1 — + octave
2 — + octave + fifth
3 — + 2 octaves
4 — + 2 octaves + major third

B: creating waveforms out of different harmonic overtones

no harmonic overtones

odd harmonic overtones, decreasing at 12dB/octave

all harmonic overtones, decreasing at 6dB/octave

odd harmonic overtones, decreasing at 6dB/octave

Figure 9.2 Harmonic overtones.

faster than if you use brand new ones. And if you sing a crescendo, you can hear your vocal cords producing more and more overtones. The sound will not only become louder, but richer as well. The effect is surprisingly similar to electronic overdrive, so no wonder that distortion is so popular as a means to liven up boring sounds.

The Weakest Link

It's different for every device what exactly happens when you send it into overdrive, but in every case, the weakest link in the circuit is no longer able to transfer the signal linearly. Sometimes it's the transformers, at other times the amplifiers, inductors (coils), magnetic tape, AD converters or your software's digital ceiling. The principle is always the same: the signal wants to go upward in a straight line, but the circuit can't keep up and bends the line downward.

The shape of this bend can be visualized in a transfer curve, and it determines the sound of the distortion (see Figure 9.3). Distortion adds overtones—frequencies that weren't there before—to the signal. Just think of what happens when you keep pushing a sine wave further into over-drive: slowly but steadily, it will start to look like a square wave. On a spectrum analyzer, a square wave will look very different, as it has an infinitely long series of odd overtones, which a sine wave doesn't have.

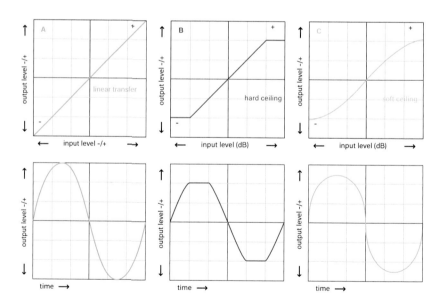

Figure 9.3 The effects of different transfer curves on a sine wave. A sharp angle causes a lot of high-order, dissonant overtones, while a smooth transition results in more low-order, relatively consonant overtones.

Just Like in Real Life

When you listen carefully, you'll notice that electronically generated distortion doesn't necessarily have a natural sound. Still, many listeners subconsciously accept it as normal, even in acoustic music. This is because harmonic distortion also occurs naturally, due to mechanical or acoustical causes. Just think of a membrane that generates sound by vibrating, for example a drumhead. It's only elastic within a limited range, and when it can't stretch any further at the extremes of this range, the drumhead starts to hold back the vibration peaks. That's just as much distortion as overdriving a tube amplifier is. The harder you play on your drum or guitar, or the more you strain your vocal cords, the more overtones you'll hear due to harmonic distortion. Even our ears are anything but distortion-free. So distortion is all around us, and it's a sound we mainly associate with loudness. And that, of course, makes it a very useful effect in music production.

Hard and Soft

The makeup of the added overtones depends on the clipping character of the device being overdriven. If the distortion is symmetrical (if it's identical for the positive and negative half of the signal), it will only produce odd overtones. However, if the distortion of a device is asymmetrical, this will result in both even and odd overtones. Even overtones are very pleasant-sounding, as the first two are exactly one and two octaves higher than the source. This kind of distortion is very inconspicuous, but it does make the sound richer. Especially devices with a 'soft ceiling' really exploit this principle, as they add a relative high number of these consonant intervals.

The harder the curve's knee, the more high-order overtones are added, which relate to the source signal at increasingly dissonant intervals. The hardest knee is digital clipping, as a digital system can react perfectly linearly up until its ceiling, but above that, it has no headroom at all. This type of distortion sounds so sharp and crackling that it's rarely seen as desirable. Some analog devices come close to this kind of hard clipping, but the diodes used—in fuzz pedals, for example—have a much larger transition region than an AD converter.

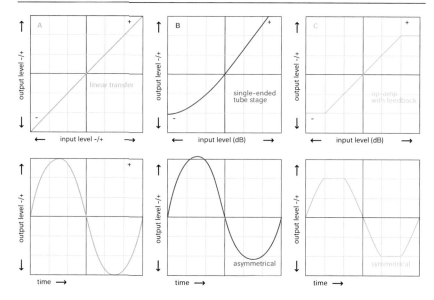

Figure 9.4 The difference between symmetrical and asymmetrical distortion. Asymmetrical means that positive and negative voltages don't distort the same way. One half of the wave distorts more than the other, which results in pleasant-sounding, even overtones. Some instruments (such as trumpets) already produce asymmetrical waveforms by themselves.

9.1.2 Distortion in a Mix

The main reason for using distortion is to suggest loudness, but there are many more reasons. Distortion can be a good alternative to EQ and compression, because, like these processes, it affects the dynamics and frequency spectrum of a signal. You can make details come to the surface, enrich the frequency spectrum, attenuate peaks, make sounds decay longer, and so on. All of this can be done with EQ and compression as well, but distortion still sounds completely different. EQ only affects already existing frequencies, while distortion actually adds new ones. An equalizer can influence the frequency spectrum without changing the dynamics, but with distortion this is impossible.

Conversely, compression can affect the dynamics without radically changing the frequency spectrum, which is also impossible with distortion. On top of this, compression has a time response (it fluctuates with the music), while distortion starts and stops instantaneously. Compression can be used over a wide dynamic range, while most distortion only sounds good within a small dynamic window. Above the window's boundary the distortion will quickly become extreme, while below there's usually not so much going on. That's why the combination of compression and distortion works very well: the compressor comes first in the chain, keeping the dynamics of the signal you want to distort within the range where the distortion sounds nice (see Figure 9.6).

Types of Distortion

Fuzz, clipping and overdrive are only a few terms to describe what a certain distortion effect does. For example, the fuzz category can be subdivided into countless variations, based on the type of diode they use. The abundance of names might be clarifying to the connoisseur, but is possibly confusing as well, as the terms can mean something else to different people. Technically speaking, any form of harmonic distortion is overdrive: you overdrive a device until it distorts. The differences are in the hardness of the ceiling and the amount of distortion. These are the most commonly used names, from soft to hard: (tape) saturation, drive, crunch, overdrive, distortion, high gain distortion, fuzz, (hard) clipping, bit crushing. The first part of this list refers to overdriven analog circuits or emulations of these, while bit crushing sounds completely different. This effect occurs when you reduce the bit depth of your audio files to very low values without dithering, for example from 24- to 6-bit. The remaining dynamic range will be very low, which creates quantization distortion: extremely aggressive, crackling distortion.

The range of sounds you can make with different types of distortion is so incredibly wide that it's almost incomprehensible. A fun and educational way to explore some possibilities is by trying out Decapitator by Soundtoys (or something similar). This plugin contains models of a number of different circuits. You'll instantly hear the effects of various series of overtones: some almost seem to add an octave below the fundamental, while others place much more emphasis on the high frequencies. On top of this, you can choose to focus the distortion more on the low or high end, which allows you to create all kinds of different sounds.

Frequency Dependence

If you've ever sent a loud sine sweep through your speakers, you'll know that it's harder for the average speaker to reproduce the lows without distortion than the highs. In order to reproduce low frequencies, the excursion of the cone has to be relatively large. As a result, it will be stretched to its limit earlier. This kind of frequency-dependent distortion occurs in more places: transformers, for example, also distort more on low than on high frequencies. And it gets even crazier, because they distort the most on extremely strong and weak signals. Many old-fashioned tube microphones also distort more on low than on high frequencies. And recording on tape comes with a real cocktail of non-linear mechanisms. The setting of the recorder and the type of tape greatly affect the way different frequencies distort. This kind of frequency dependence can be very useful musically: transformers and tape recorders make the lows both warmer and deeper, without being recognizable as distortion.

Figure 9.5 Analog tape.

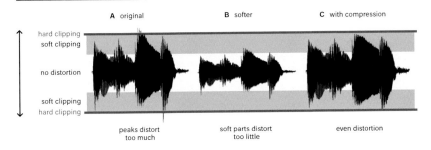

Figure 9.6 The range in which overdriving a device sounds nice (orange) is often not so wide. If you exceed the red line (hard clipping), the distortion will become too extreme, but if you stay in the yellow area, the effect is too small. By compressing the input signal first, you can keep it within the orange zone as much as possible.

Restricting

What's so addictive about distortion is that it seems to make everything more energetic, thick and fat. But therein lies the problem, because if you use a lot of distortion, the sounds you create will be so dense that other sounds will have a hard time cutting through. This is what makes high-gain guitars or sawtooth synths so hard to mix: these sounds have such a dense frequency spectrum—and often such even dynamics—that they can end up covering the entire mix like a blanket. This can be avoided by using filters to carefully restrict the area within which these sounds are permitted. The speaker of a guitar amp already barely reproduces the very high frequencies, but an additional low-pass filter can still create a lot more room in the mix. The effect of this is that you retain the lower overtones (which have a strong harmonic relationship with the signal), while attenuating the higher, dissonant distortion (which isn't very powerful musically). Because of this, distortion and low-pass filters are a popular combination.

Focusing

Instead of using filters to process the signal after it's been distorted, you can also focus the distortion more precisely on a specific frequency range. This can be useful when you can't allow an instrument to take up much space, but you still want to put some emphasis on it. Making bass instruments audible in a busy mix is a common application of this. Putting distortion on the entire instrument will produce a lot of density in the octave directly above the fundamental of the bass. Usually, that's exactly where the fundamentals of the harmonic instruments are, so this range will quickly turn into a muddy cluster of sounds. This problem can be

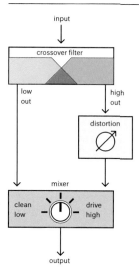

Figure 9.7 With a crossover filter (which you can also create yourself, by using two separate filters) you can keep an instrument's low end distortion-free. This way, the mix will still sound open, while the distortion makes the midrange cut through the mix more.

solved by dipping the distorted bass between 150 and 300 Hz, but there's another option as well. If you split the bass into a low and a high part, you can choose to only distort the high part. This way, you'll mainly generate more overtones on the overtones of the bass. As a result, you'll hear these tones more clearly, and the instrument will cut through the mix more. The fundamentals remain undistorted, so the bass won't push other instruments away. This is exactly the reason why a lot of producers combine a clean DI signal with a distorted amplifier signal, as this will give them both 'smooth' lows and a 'biting' midrange. Some distortion effects have built-in crossover filters (or pre-emphasis filters) to make this type of frequency-focused distortion easy.

Blending

By definition, distortion evens out the peaks of your signal. Usually that's not your main reason for using it, as you're mostly interested in the sound of the distortion. Sometimes you wish you could capitalize on this timbre, without having to flatten everything. This is because mixes tend to sound 'overcooked,' rhythmically weak and generally undefined if there aren't enough peaks left. You can solve this problem by blending the distortion and the original signal together. This way, you can make a combination of the original peaks and the added timbre. When you use this technique, make sure that the original and the distorted version add up in phase.

This is not always easy, as many distortion effects also contain filters that can cause phase shifts. In these cases, all-pass filters can help to keep the low end in phase (see section 7.4). Parallel distortion is the ideal way to give sounds more definition and character, without turning the mix into a tiring and flat listening experience.

Distortion can save a lifeless piece of music or ruin a beautiful song. The trick is to use it functionally, and to avoid using it if the music doesn't need it. Guitarists know how to play this game, since distortion is such a major part of their sound. They adjust the amount of distortion to the part they're playing, and vice versa. For example, when there's a lot of distortion, they often remove the third from the chord to keep the sound more harmonically open, plus they will emphasize the rhythmic attack a bit more. In music production, a mindset like that can help to convincingly integrate distortion into your mix. And once you manage to do that, take cover!

9.2 Re-amping: Providing Context

Re-amping is nothing more than playing a sound back through an amplifier and a speaker, and then recording it again with one or more microphones. But why would you want to do that with instruments other than directly recorded guitars? Apparently, what's added to a sound on its way from a speaker to a microphone is hard to achieve in a different way: depth and perspective. Of course you can also record sounds this way from the start, but if you don't have this option, or if you're working with electronic sources, re-amping is a pretty close approximation of this process. On top of this, there's a lot to be gained in terms of sound texture, as amplifiers and speakers can be pushed into overdrive. And a distorted sound that's recorded via a speaker in a space with its own acoustics is usually easier to place in the mix than a sound with distortion directly applied to it. It won't be as dominant in the mix, and it automatically has a clear place in terms of depth—which you can easily manipulate by moving the microphone(s). So plenty of reasons to warm up those tubes!

9.2.1 How it Works

The re-amping chain that you put together should of course match the effect you have in mind. A small speaker can work well to 'focus' the sound, while a large woofer can remove all kinds of unwanted high-frequency noises. When you add distortion, the small speaker will sound even brighter and smaller, while the large speaker will seem even fuller and thicker. The resonance of a loudspeaker within its enclosure always gives the sound an emphasis around a certain frequency. You can see this as the system's lower limit. If this frequency fits the instrument you're running through it, it will seem to have more body. This is why guitarists value the diameter of their speakers so much. If the speakers are too

Impedance

Guitar amps are designed to deal with the (generally) high output impedance of electric guitars. As a result, they have a much higher input impedance than the line inputs of studio equipment. So if you want to record a guitar directly, it's a good idea to use a DI box or hi-Z input. This way, the guitar's frequencies will be transferred in the same proportions as when you send them to an amplifier. If you don't do this, you're bound to lose the high end. And if you later want to run this recording through a guitar amplifier, you can apply the same principle in reverse by using a re-amp box to increase the impedance (and attenuate the signal level) before it goes into the amplifier. An additional advantage of a re-amp box is that it usually has a transformer that breaks ground loops, thus preventing hum.

This doesn't mean there's a law that says impedance should always be matched. If you don't need to exactly reproduce the sound of a particular guitar/amp combination, DIs and re-amp boxes are often unnecessary. I sometimes use stompboxes as an effect while mixing, directly connected to the converters. I keep the converter levels low, and after the effects I (digitally) add some gain to match the gain structure of my mix. If there are any small timbre differences due to an impedance mismatch, I can usually live with that. It varies per effect how it turns out, but if it sounds good, it is good.

big, their response will be too slow for some playing styles, but if they're too small, their emphasis will be too high up the frequency range to add enough weight to certain parts.

At the same time, one-way systems like guitar amplifiers also have an upper limit because they don't have tweeters. The larger the speakers, the sooner you'll lose the high end. This loss can work well to temper dissonant overtones caused by extreme distortion. Of course you're not restricted to guitar amps, as all speakers can be used for re-amping, and you can even hook up multiple speakers to create all kinds of spatial effects.

Speaker Placement

If you want to make a good recording, you have to make sure the speaker sounds good in the recording room. This process is a lot like positioning your studio monitors or choosing the best place for your drum kit. The speaker's interaction with the space depends mostly on its own directivity and the room's geometry. The direction in which you point it determines the sound, level and arrival time of the early reflections you'll hear in your recording. The placement of the speaker is particularly critical for the lower part of the frequency response, which is greatly affected by room modes. If placement and angle alone don't get you there, you can also use acoustic aids like gobos or diffusors to fine-tune the sound. Getting to know your space by experimenting with placement is the only way to really get a grip on the process, as there are too many mechanisms working simultaneously to be able to provide a formula for good placement. Besides this, 'good' is subjective and strongly dependent on the musical context (more information on acoustics and speaker placement can be found in Chapter 17).

Listening to the effect of placement in the recording space is extremely important, but it's not always possible. With loud (guitar) amplifiers, there's a risk of hearing loss. And even if things don't get that ear-splittingly loud, your ears will quickly get used to the high sound levels, and two minutes later you won't be able to say much about the sound of your recording in the control room. It's better to stay in the control room while you make an assessment of the problem and come up with a solution, then make adjustments in the recording room (in silence), and finally return to the control room to evaluate whether your adjustments worked. For those who are good with robotics: a remote-controlled guitar amp platform or a microphone placement robot are handy gadgets to have. An assistant you can instruct via isolated headphones is also an option, of course.

Distortion in Stages

In a typical 'guitar chain,' the sound can be distorted in a number of places. Stompboxes, preamps, power amplifiers and speakers can all be pushed to their limits. But stacking distortion on distortion on distortion doesn't necessarily have a positive effect on the definition of the sound. So if you mainly want to hear the distortion of the power amp and the speaker, it's best to set the preamp gain a bit lower, and vice versa. Varying this multi-stage gain structure is something guitarists can endlessly tinker with, as it greatly affects the dynamics, timbre and harmonic density of the result.

Save Your Ears

Pushing the tubes of a guitar amplifier into overdrive can sound great, but the resulting sound levels are no picnic. Fortunately, there are several ways to limit this inconvenience. An isolation cabinet containing a speaker and a microphone of choice is closest to the real thing. The only downside to this is that the cabinet has internal acoustics, which can never be completely damped. As a result, these solutions sometimes literally sound a bit 'boxy.' Another approach is to use a so-called power soak to attenuate the signal after overdriving the power stage. This will make the speaker work less hard, while the amplifier is still distorted. The drawback to this is that you reduce the distortion the speaker itself introduces at high levels, so it still doesn't sound exactly the same. The most extreme solution is a dummy load (a heavy resistor that can fully absorb the power of the amplifier) with a line output. This allows you to tap the amplifier signal directly, without having to put a speaker in motion at all. The problem is that besides the distortion of the speaker, you also miss its frequency response this way. That's why many dummy loads come with a low-pass filter, which simulates the frequency response of a woofer.

Microphone Placement

If you've never recorded electric guitars before and you're just starting to read up on the different methods engineers use for this, the recipes you come across can seem extremely detailed. 'Place a type X microphone at five centimeters from the speaker, two centimeters away from the cone's axis, at a 30-degree angle,' and so on. The reason for this is simple: this close to the sound source, tiny changes can make huge differences.

Speakers and microphones are definitely not technically perfectly functioning inventions. The smoothed-out frequency response graphs that manufacturers like to advertise with are only true for sound that's perfectly on axis with the speaker or microphone. If you use them at an angle, and not in an anechoic chamber but in a room with acoustics, the frequency response can change dramatically. This is why speaker and microphone placement is a lot like turning the knobs of an equalizer, only with less predictable results. In general, you can say that speakers and microphones radiate and pick up low-frequency sound more omnidirectionally, and that they become more and more directional as the frequencies get higher.

You can also use multiple microphones, of course. The combinations of different microphones and distances are endless, but a single microphone in the right place always produces a bigger sound than a mix of three microphones that don't work together well. A good rule of thumb is to only add extra microphones if the sound they pick up is actually different, and not just a variation on the same theme. For example, stereo room mics placed a few meters away from a single close mic can add a lot of extra width and depth. Due to the difference in distance and the relatively high level of diffuse reflections picked up by these microphones, they add much more to the close mic than they take away through phase interaction. If you intend to blend your re-amped recording with the original source, it's important to precisely correct any delay and phase shifts introduced by the re-amp chain. Close mics should be aligned as accurately as possible with the original signal, while room mics usually benefit from separation: a little delay compared with the original goes a long way (see sections 7.2 and 7.3 as well).

A: flat frequency response

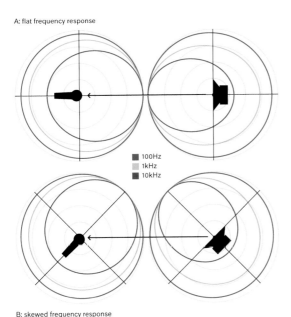

100Hz
1kHz
10kHz

B: skewed frequency response

Figure 9.8 The frequency response of speakers and microphones depends on the angle at which you place them. The bottom setup will sound a lot less bright than the top one.

9.2.2 Re-amping in a Mix

Adding Air

A common mixing problem is the depth placement of electronic sources or directly recorded instruments. You can add reverb to these instruments, but the resulting effect will often be a very direct-sounding source with the reverb trailing behind it. In cases like these, re-amping can add a lot of realism, especially if you precisely adjust the speaker and microphone setup to where you want to place the sound in the mix. Figure 9.9 shows how you can make a very direct-sounding violin 'breathe' a bit more. If the speaker that simulates the violin isn't aimed straight at the microphones, but at an angle (just like a violin at a concert is at an angle with the audience), it will already sound a lot less direct. The microphones are placed at some distance to the side, where they pick up a realistic stereo image (in terms of width and depth) of the instrument and its reflections. Naturally, this method can only work in a space with good acoustics: in a room with a low ceiling you'll have the same problem as with a real violin, so the distant microphones won't sound very good.

Stereo Movement

It's not easy to turn a mono source into something that fills the entire width of the stereo image in a meaningful way. If it mainly has to be impressive—and realism is less important—you can use multiple speakers to reproduce the same sound source, and make them all sound different. There are many ways to create these sound changes, for instance by using two guitar amps, one with overdrive, chorus and delay, and the other

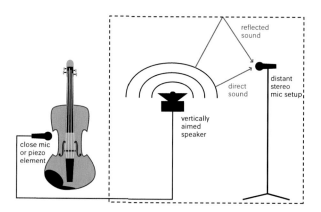

Figure 9.9 Just like an instrument, you can make a speaker radiate in one particular direction, thus changing the sound of the recording.

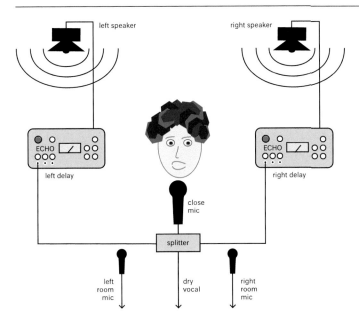

left speaker

right speaker

ECHO

ECHO

left delay

right delay

close
mic

splitter

left
room
mic

dry
vocal

right
room
mic

Figure 9.10 Recording vocals with three microphones and live stereo delays in the same space will result in a completely different mix than when you add the delay afterwards. The spill makes everything more intertwined.

clean. Aim one amp toward the corner of the space with the microphone a bit further away, and place the other one in the middle of the space with a close mic. Pan the resulting tracks to the left and right and voilà: you've created an impressively wide sound that's much more complex than the average stereo synthesis plugin.

Combining Acoustics and Amplification

The main problem with combining electronic and acoustic sounds is that it's difficult to place them in the same world. Of course, the ideal solution would be to literally place them in the same world. Instead of programming an electronic beat first and separately recording all kinds of acoustic additions, you can also play this beat in the recording space while you record the piano. With a couple of room mics, it will immediately sound a lot more convincing than when you use electronic reverb. Figure 9.11 shows how to complement a drum kit with a PA in the recording room. When it's amplified, the acoustic snare drum will be louder in the room mics. As a result, the drum kit will seem a lot bigger than when you only boost the close mic of the snare in the mix. And by not only recording the triggered electronic drum sounds directly, but also

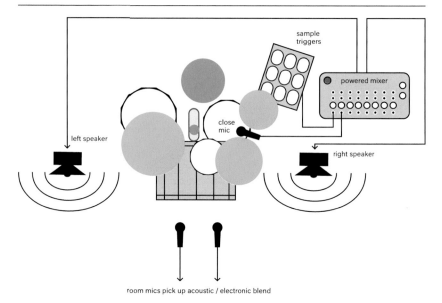

Figure 9.11 Combining acoustic drums with electronic elements is a lot more convincing if you amplify the electronics live and then record everything at once with the room microphones.

amplifying them in the space, they will automatically blend with the acoustic drums more.

9.3 Modulation: Putting Things in Motion

If you've ever worked with a modular synthesizer, you'll know the power of modulation. Influencing one parameter with another helps to put a sound in motion and give it complexity. With effects, it's the same: when they move, they attract more attention than when they remain static. Examples of this kind of motion are:

- motion in pitch (vibrato and chorus);
- motion in amplitude (tremolo and envelope manipulations);
- motion in spectral composition (filtering, flanging and phasing).

The advantage of sounds that change with time is that our brain pays more attention to them, which makes them stand out more in the mix as a whole. Because of this, modulation can also help to reduce masking: an instrument that's easier to hear can be turned down, so it won't get in the way of the other instruments so much.

9.3.1 How it Works

Vibrato, Chorus and Flanging

Pitch modulation works just like a passing fire truck. The Doppler effect that causes the pitch to drop is the result of the fire truck first moving toward you and then away from you. So actually it's distance modulation that causes this fluctuation (see Figure 9.12). Delay is a perfect tool for simulating distance: the longer the delay time, the greater the distance. If you change the delay time during playback, you'll hear a similar pitch bend. (Or you'll hear a lot of rattle, as some digital delays don't make smooth movements when you adjust the timing. This effect is called zipper noise, and it's usually undesirable.)

With the delay set to 100 percent wet, you can make Doppler-like effects by modulating the delay time. The resulting pitch fluctuations are mainly interesting as a special effect: think of imitating a wobbly tape recorder, making musically timed pitch sweeps, or simulating vibrato. You often have to use quite a lot of modulation to make the effect stand out, so much that it will sometimes sound a bit out of tune. This changes when you blend some of the raw signal with the delay. Then, even the

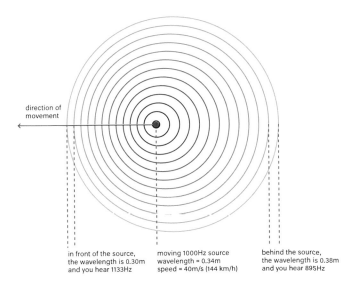

direction of movement

in front of the source,
the wavelength is 0.30m
and you hear 1133Hz

moving 1000Hz source
wavelength = 0.34m
speed = 40m/s (144 km/h)

behind the source,
the wavelength is 0.38m
and you hear 895Hz

Figure 9.12 The Doppler effect. The movement of a sound source results in greater distances between the pressure variations behind the source (longer wavelength = lower pitch), and shorter distances in front of it (higher pitch).

slightest pitch fluctuation will cause a huge change in sound. This is the result of interference: the original and the delayed signal are very similar, so they partly amplify and attenuate each other when you combine them. With average delay times—somewhere between 10 and 40 ms—this principle produces a chorus-like effect. It's a lot like double tracking a part, and it makes the sound richer.

With very short delay times, the interference gets stronger, resulting in comb filters with regularly (harmonically) distributed peaks and dips. The location of these peaks and dips keeps changing with the delay time. You can make the peaks even more extreme by adding feedback to the delay, which produces a whistling sound. This effect is called flanging, and it's nothing more than a chorus effect with very short delay times. If you create this flanging by using tape recorders as delays (the way it was once discovered), the delay times can even become negative. This is the result of playing the same recording on two tape recorders, which you both manipulate in terms of playback speed. The copy can then end up both before and after the original, and this kind of 'through-zero' flanging undeniably has a beautiful character. Digitally, this is easy to imitate by delaying the original as well, and if you then also add tape simulation to both signal paths, you'll get a pretty good approximation of this sonic character. Due to their imperfections (skewed phase response, fluctuating playback speed, distortion), tape recorders add an unpredictable, lively dimension to the sound when used as a parallel effect. For example, you can double track parts more convincingly with a tape recorder than with a chorus effect, as if they were really played twice. There's even a name for this process: automatic double tracking (ADT). Many chorus effects sound too 'perfect,' so the duplication doesn't feel human.

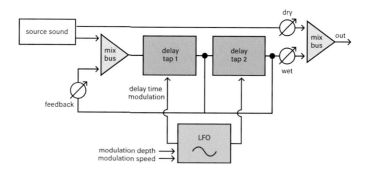

Figure 9.13 Chorus and flanging are nothing more than modulated delays that you blend with the original. You can make the sound more complex by using more delays (taps). Adding feedback enhances the effect.

Natural Modulation

The method invented in the psychedelic sixties of having microphones (or musicians) spin around while suspended from cables
is a great way to accomplish acoustic pitch modulation, but this
effect is also found in less adventurous situations. For example, a
violin section playing in unison is similar to an irregularly modulating 18-tap chorus effect, each delay tap with its own panning.
On top of this, violinists can play with vibrato, and they usually
don't hold their instruments still either. So there are more pitch
fluctuations in the acoustic world than you might expect. It's just
that they aren't nearly as regular and predictable as electronic
effects. As a result, they don't attract attention, but unconsciously
they do contribute a lot to your sense of size and depth. For many
mixers, this is reason enough to apply chorus in subtle amounts.
To a certain extent, it can increase your perception of size.

Phasing and Filtering

Flanging and chorus effects use a delay for the overall signal, creating a
comb filter with regularly distributed peaks and dips. If you introduce
a phase shift in the parallel path instead of a delay, you've created a
phaser that can also produce irregularly distributed peaks and dips.
A phase shift can be seen as a frequency-dependent delay, as each
frequency gets a different delay time, courtesy of an all-pass filter. The
shift is 180 degrees around the filter's cutoff frequency, resulting in
attenuation when the signal is combined with the original, while else-
where in the spectrum amplification occurs. Unlike a flanger, a phaser
doesn't produce an endless series of peaks and dips, but only a few of
them, depending on the number of all-pass filters (also called stages) used
by the phaser. The cutoff frequencies of these filters can be modulated
with an LFO, which constantly changes the frequency where the peaks
and dips occur. A phaser can also use feedback to make its response even
stronger.

So phasing is actually nothing more than the constant moving of a
number of notch filters. But of course you can think of many more types
of filters that can create interesting effects when you modulate them.
A simple example is a low-pass filter that filters out the hi-hat spill
from a snare drum microphone, except when the snare drum triggers

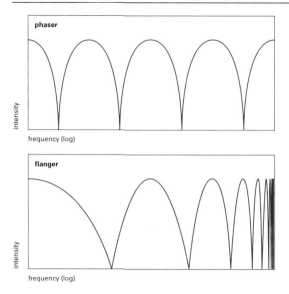

Figure 9.14 The difference between the (static) frequency response of a phaser and a flanger.

the filter, causing the cutoff frequency to shift upward. Or two filter banks, one for the left and one for the right channel, which remove and pass opposite frequency ranges. Add some subtle modulation to it, and you'll have a great setup for generating a wide stereo image from a mono source.

The most fun example of a modulating filter is probably the talk box, a small speaker that sends the source sound through a plastic tube and into the mouth of the player, who creates a modulating acoustic filter by changing the shape of the mouth. The filtered sound is then picked up by a microphone.

Tremolo and Ring Modulation

Turning a sound up or down is the most basic manipulation there is, but you can still create very powerful effects with it. Tremolo is simply the result of modulating the sound level with a low-frequency oscillator (LFO). This LFO generates a periodic signal (a waveform) that drives the modulation. Of course, the tempo of the modulation (the frequency of the wave) can be synchronized with the music, and the shape of the LFO's waveform has a very strong effect on the feel of the final result (see the box 'Modulation Sources'). If you keep increasing the frequency of the LFO, you'll slowly turn the tremolo into a ring modulator, as this creates sum and difference tones between the source signal and the modulator. This effect is also called amplitude modulation, and it can produce beautifully dissonant sci-fi sounds.

Rotary Speaker

The rotary speaker or Leslie speaker (named after its inventor) is mainly known for the recognizable electronic organ sounds heard on countless records. The bulky cabinet contains a rotating sound reflector connected to a woofer for the low frequencies, and an independently rotating horn tweeter for the high frequencies. Both rotation speeds can be adjusted separately. Besides a source of backaches for roadies, this double rotation system also produces an extremely lively sound, in which vibrato, tremolo and a modulated filter effect can be heard, independently for the low and high frequencies. Add the overdrive of a tube amplifier to this, and you have a very complexly responding sound machine. The Leslie speaker has been imitated a lot, but in its original form it still sounds unique.

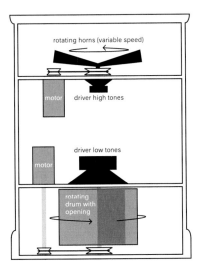

Figure 9.15 Leslie speaker cabinet.

9.3.2 Modulation in a Mix

Width

One of the main reasons for using modulation effects is to create stereo width without the need for large amounts of reverb. By manipulating the left and right channel separately, with different modulation speeds, opposing movements, (inverse) feedback between left and right or different frequency characteristics, you can give the sound more dimension. All of the modulation effects mentioned in this section can be used this way, but chorus is the most obvious choice. It's less conspicuous than flanging or phasing, and it can make the sound a lot thicker and bigger.

Attention

A car that's rushing by will attract your attention more than a stationary one, and in a mix, moving sounds also seem more important than static ones. You can exploit this principle by putting sounds in motion if they need to draw more attention. The most obvious way to do this is through panning automation. Besides a ready-made auto panner, you can also use two separate tremolos for the left and right channels to achieve this. This way, you can create very complex and unpredictable movements. But if you overdo it by making the movements too fast, you will lose definition— and therefore attention—in your mix. Slow, handmade sweeps with panners, phasers, flangers or filters often give the best results.

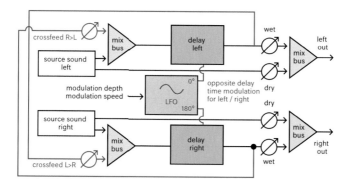

Figure 9.16 Stereo widening can be easily achieved by modulating the left and right halves of an effect in opposite directions. If you then also add feedback between the left and right channels (crossfeed), the combined sound will become even wider.

Distinction

Another way to utilize the extra attention that moving sounds attract is by turning these sounds down in your mix. Thanks to their distinctive texture, modulated sounds can be musically effective at a lower mixing level than static ones. This is ideal for doubled parts and other additional elements that take up too much space, because the more softly you can place these extras in your mix, the more room you'll have left for the important things, and the more powerful your final mix will be. Reverb and delay can also be seen as extras: if you put chorus or phasing on the reverb and delay returns, you can usually turn them down a bit, while achieving a similar perception of size.

Rounding Off

An application of chorus, flanging or phasing that you might not think of right away is making sharp sounds like metal percussion milder. Tambourines or other piercing percussion sounds that stubbornly stay in the front of the mix, no matter how much reverb you put on them, can be pushed back quite far this way. Comb filtering and phase distortion make the transients less sharp: by spreading them out a bit in time, they become less direct-sounding. This approach will sometimes sound more pleasant than a low-pass filter, which also makes the transients milder, but often ends up muffling the sound as a whole.

Modulation Sources

Just as important as the sound of the effect you use is the source that controls it. This can make all the difference between an effect that sounds predictable and 'tacked on,' and an effect that blends with the source and appeals to the imagination. The most basic modulator, a constant oscillator, can be given different rhythmic functions by changing the waveform. For example, a sawtooth has a sharp emphasis, while a sine has a more spread-out pulse. The phase of the waveform determines the position of the rhythmic emphasis: on the beat or between the beats. But you don't need to stick to the tempo, and you can even go as far as modulating the tempo of the modulator itself. This way, you can create alternately rushing and dragging movements. If this is still too predictable, you can also consider using a random generator (noise) as a modulation source. Especially if you cut some of its highs, you can introduce very unpredictable, slow movements with it, which sometimes sound very 'human.'

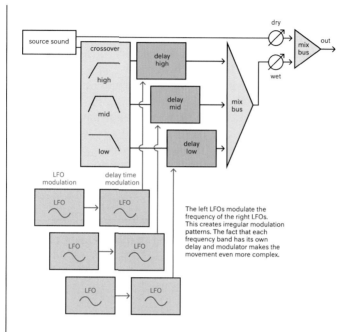

Figure 9.17 A way to make chorus sound less regular and predictable.

On the other hand, if the sound should be more recognizable, you can use a step sequencer as a modulator. This creates rhythmic patterns in the sound that go much further than simply repeating the same movement like an oscillator does. This way, you can use a filter to turn long tones into arpeggio-like sounds, without changing the pitch. One way to make sure the movements made by your effects are related to the music is by using the music itself as the modulation source. Auto-wah is a well-known example of this: the louder you play, the higher the cutoff frequency of the filter. But with this kind of envelope modulation, you don't necessarily have to use the signal you're manipulating as the source: you can also use the level of the drums to modulate a phaser on the bass.

The danger of all automatic modulation constructions is that they easily end up sounding contrived and drawing more attention than necessary. When you think about it, the trick in music is to use movement sparingly and to not keep all the elements constantly in motion. The best—and most fun way—to do this is by appointing yourself as the modulator. A filter sweep or 'jet flange' that precisely follows the song's structure is made quickest by hand. This way, you'll automatically end up with effects that are only active during the musical moments they're meant to affect. These are much more powerful than a tremolo that keeps rippling for minutes on end.

9.4 Pitch Manipulation: Creating What Wasn't There

If you've ever played an LP record at 45, or even better, at 78 rpm, you know what pitch shifting is. The mechanism is very simple: you play something faster or slower than it was recorded, and the pitch shifts up or down. You can do this digitally as well: the sample rate then becomes your recording speed. If you tell your computer to play 96,000 samples per second, but the file you're playing was recorded at 48,000 samples per second, it will sound an octave higher. And be twice as short! Which is exactly the problem of pitch shifting, because in many cases you don't want to change the length of a sound, only its pitch.

9.4.1 How it Works

Time Stretching

If you want to raise or lower the pitch of a sound, but maintain its duration, you'll need to lengthen or shorten it. Technically, this simply means repeating or cutting out tiny pieces. But the trick is to do this inaudibly, because if you've ever made a cut at a random point in your audio, you'll know that an interruption of a waveform almost always causes a loud click or other noise. Crossfading is a good way to solve this problem, but in the middle of a long note it's still a challenge. It's important to keep a close eye on the phase of the waveform: if you remove or add one or more entire periods (repetitions in a waveform), you'll hardly notice it, because the crossfaded parts will then be in phase.

If the sound isn't periodic, because it's undergoing a change of pitch or because it contains noise sounds (transients are a good example of this), it's much more difficult to lengthen or shorten it. Therefore, a smart pitch shifter studies the signal to see which parts are periodic, and which ones only occur once. The latter won't be time-corrected by the device, only the periodic parts. For instance, in a spoken sentence, the vowels are stretched or shortened, while the consonants are manipulated much less severely.

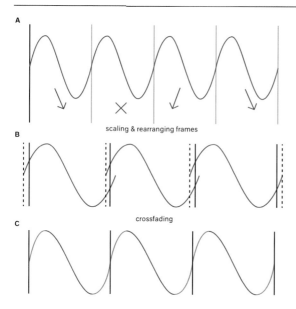

Figure 9.18 Lowering a tone through pitch shifting with time correction. The system divides the original wave into a whole number of periods (A). Then, at the new pitch, it tries to create a continuous wave of the same length which also has a whole number of periods. In this case, the system cuts out one period and rearranges the remaining ones for an as-close-as-possible approximation (B). But because it's still not perfect, the difference is corrected through interpolation (crossfades). This creates some distortion (the yellow sections in C).

FFT

An alternative method for pitch shifting is the so-called phase vocoder. This principle uses FFT (see section 9.6 for more information on this) to convert a sound to the frequency domain. This way, you can see the frequency components (and their phases) that make up a signal. This information can be manipulated, which allows you to easily perform time and pitch adjustments without interruptions or phase cancellations in the signal. After all, you'll control the phase and pitch of each of the frequency components in every successive piece of audio (these pieces are called bins in the frequency domain). This way, you can make sure there's a smooth transition from one bin to the next when you copy or remove a few.

Some pitch shifters use this method for time correction, and then resampling for pitch adjustment. Other pitch shifters also adjust the pitch in the frequency domain, which creates a lot of extra possibilities, like shifting the pitch of individual frequency components or copying them to generate harmonies. It's even possible to isolate individual notes and their overtones, and shift them compared to other notes (polyphonic pitch correction). Still, it's not easy to do this without side effects. This is due to both the difficulty of isolating the right frequency components and the inherent resolution problems of FFT. As soon as you want to move something by anything in between a whole number of bins, you'll need to interpolate between different bins, which produces all kinds of by-products.

Formants

Usually, you want to use pitch shifting to create the illusion that the same instrument produces a different note. To make this work, you want to simulate that the strings vibrate at a different frequency, while the soundbox stays the same. This is not easy: with standard pitch shifting, the resonances of the soundbox shift along with the new pitch. The human voice also has a soundbox with resonances, which are called formants. The frequency of these resonances is quite constant; all you can do is modulate them a bit by shaping your mouth. The effect of these formants is that they amplify some of the frequencies produced by your vocal cords more than others.

For a natural-sounding pitch shift, you want to keep this 'EQ effect' at the same frequency, while adjusting the frequency of the vocal cords. There are pitch shifters that try to distinguish the influence of the formant filter from the source of the vibration. Then the fundamental and its overtones are shifted to a new pitch, and an estimation of the original formant filter is put on top of it. This principle is called formant correction, and it can greatly expand the usable range of a pitch shifter.

Imperfections

The signals that are the hardest to shift contain few periodic waves: percussion, for example. In these cases, the side effects create a granular/grainy sound, which is sometimes exactly the effect you're looking for.

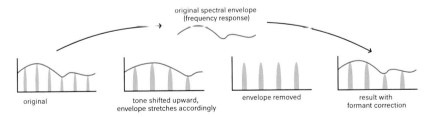

Figure 9.19 A schematic view of formant correction. The original filter character of the soundbox is retained as much as possible, while the pitch shifts upward.

In old-fashioned pitch shifters, this effect was even stronger, because the grain size—the size of the chunks the audio is divided into before it's time-corrected—wasn't perfectly matched to the waveform. This was inevitable, as these pitch shifters had to perform their task almost in real time. There was no time for offline arithmetic. Nowadays, things are better of course, but due to its 'drawbacks,' the old technology is still a popular tool for making grainy textures. Especially if you also add delay and feedback, you can create unreal, mysterious worlds with it. Feedback can be used to make pitch bends: a shift down that you send back to the input via a delay is shifted down even further, and so on. And some processors even allow you to set the grain size yourself or play the grains backwards: instant vintage sci-fi!

9.4.2 Pitch Shifting in a Mix

Supplementing

Pitch shifting can be used for many other things besides creating cartoon effects. For example, it can be very useful if the low end of a song's arrangement doesn't work so well. A section in which the bass plays high notes can be EQ'd all you want, but this won't give you a full low end. What does work is generating an octave below the bass, as this will make the mix feel bigger without attracting too much attention. An easy way to do this is by using a subharmonic generator that's specifically developed for this purpose. But it also works great to copy the entire bass, shift it an octave down and use a filter with a very steep slope to attenuate the area above the lower octave. Just make sure you don't get any phase problems when you blend the copy with the original. You can also replace entire parts with octavated versions, but this is bound to stand out as a manipulation. A blend usually works better. Just try to shift a drum room mic down an octave and blend it with the dry drums. Ka-boommmm!

Pitch shifting is also a great tool for harmonizing melodies. Some pitch shifters even let you set a specific key, so they can automatically generate suitable polyphonic harmonies. It's only when you place them loudly in the mix that they will start to sound fake, plus the timing is so neatly synced with the original that it won't sound like a separate recording. It's better to select a spare take of the lead vocal and use pitch correction software to manually turn it into any harmony you want. It's a bit more work, but the result can sound very realistic. For melodies it can also work to simply add an octavated version of the original, though not as a convincing extra part, as the shift would be too extreme for this. But as a supplement or special effect it can sound great. Just run an octave above the lead vocal through a reverb that's set to 100 percent wet, and use it as the reverb for the original.

Varispeed

You wouldn't expect it, but even in mastering studios it's not uncommon to use pitch manipulation. Usually, the purpose is not to change the pitch of the music, but the tempo. Playing a track faster can give it a more energetic and tight feel, and make it more suitable as a single or turn it into a dance floor smash (all DJs know this trick, and in the club, pitch control is used a lot for this purpose). The reason for not using time stretching to make tempo corrections is that playing a song at a different pitch doesn't affect the sound quality as much (in case of minor adjustments). But if you ever want to play along with the master in A = 440 Hz tuning, it's not an option of course!

During recording, it can also be useful to temporarily play your track faster or slower. At half speed (a standard feature on most tape recorders), you can record fast-paced parts with extreme precision. Of course, this means you'll have to play the part an octave lower than usual, unless the pitched-up cartoon effect is exactly what you're going for. The opposite is also possible: recording drums at double speed will give you deep thunderclaps when the result is played at the original tempo. With drums, the recording speed doesn't even have to correspond to whole tone intervals, since drums aren't strictly tied to a specific pitch. If the sound of your drums is just a tad too light for the song you want to record, or if certain frequencies clash with the tonal instruments in the track, you can simply use varispeed (if your DAW/word clock generator supports this) to record the drums a bit faster and then listen to the result at the original tempo. Sometimes this is a faster and better solution than tuning a drum kit to a pitch that doesn't sound good due to the size of the drums. A little goes a long way with these kinds of adjustments.

Corrections

If the instruments themselves are in tune, most tuning problems can be solved with a pitch shifter. A common issue is when a grand piano or other orchestral instrument that's tuned in A = 442 Hz has to be used in an A = 440 Hz production. This can be fixed by shifting the entire instrument 7.85 cents down, so it will still be in tune with the rest of the production. Of course, this particular problem could also be solved by using the varispeed recording method. Things get more complicated if the problem isn't caused by tuning. Sometimes musicians are consistently sharp or flat, for instance when they aren't used to playing or singing with headphones, or when the monitor mix isn't good. In these cases, shifting the entire part by a few cents can sound much more natural than adjusting every individual note with correction software. But when the sharpness or flatness isn't consistent throughout the part, there's no getting around the latter method—apart from re-recording.

Go Forth and Multiply

Chorus is popular as a doubling effect, but the drawback is that the repeating modulation of the delay time (which causes the pitch to fluctuate) can be too conspicuous. It sounds like an effect, not really like multiple parts being played. With a pitch shifter, you can also create chorus-like effects, but without modulation. A pitch shifter can adjust the pitch of the entire part by a few cents (100 cents equals one semitone), and when you blend the result with the original, you'll hear a chorus-like thickening of the sound, but no recognizable modulation. This effect (commonly known as micropitch) was made famous by the Eventide Harmonizers, and Figure 9.20 shows how it can be used to make a mono source sound wider and fuller. When used this way, pitch shifters can also be good stereo synthesizers. Panning a pitch-shifted copy to the opposite side of your mix sounds very different than only using delay. The change in pitch slightly disconnects the copy from the original. This reduces the effect of the copy 'pulling your ears apart' as a result of the phase shift, and it sounds more like a real double-tracked part.

Distortion

If you let go of the requirement that your pitch shifting should pass a listener unnoticed, much more extreme effects are possible than the ones we've discussed so far. The takeoff-and-landing sweeps you can make with a Digitech Whammy (a foot pedal-controlled pitch shifter) are a good example of this, and many DAWs have offline pitch shifts that can be combined with an envelope to create all kinds of sweeps, risers or tape stop effects. Sounds can get an entirely new meaning after an extreme shift (a couple of octaves down or up). Sound designers use this technique a lot: a seagull turns into a roaring dragon and a plastic bag becomes a cosmic storm. If you record at a high sample rate, you can even shift

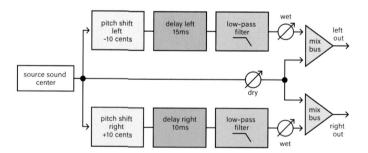

Figure 9.20 The famous micropitch effect: it gives you the width of a chorus effect, but without a recognizable repetitive modulation pattern.

ultrasound down to the human hearing range: metal percussion overtones and other things you've never heard before suddenly become usable sound material.

9.5 Triggering: Cause and Effect

Rhythm is the driving force of every mix. A good rhythm leads the listener through the music and gives contrast and energy to all the other elements. But whether this works or not mostly depends on timing. You might think this is pretty obvious, but there's a lot more to it than just the ability to keep tempo. The interpretation of 'between the beats,' the timing relationship with other instruments, the duration of the rhythmic sounds and their tuning, size, mutual balance, strength and timbre all determine the feel of the rhythm. Many parameters can still be influenced while mixing, regardless of the original performance. With faders, compressors, equalizers and reverb effects you can already manipulate a lot. But you can take things even further if you use the rhythm as the driving force to trigger special effects or completely new sounds. This opens up a world of interesting constructions that can help to put your mix in motion.

9.5.1 How it Works

The first question to answer is whether the things you trigger should respond to the loudness of the rhythm or not. For example, if you want to use the rhythm of the kick to open a gate on a bass synth, you'd prefer the gate to open on every kick, regardless of how loud it was played. To pull this off, you'll need a 'clean' trigger signal. If the kick was recorded with little spill from other instruments, you can first compress it before you let it open the gate on the bass synth. This way, at least every kick will be equally loud. And if there's too much spill from the toms, you can also EQ the kick in such a way that its own fundamental is emphasized, while those of the toms are suppressed as much as possible. This will make sure that your trigger signal is as clear as can be (see Figure 9.21).

If this fails, you can also try to trigger a sample with the uncompressed kick (compression amplifies the spill as well, which makes it harder to differentiate between the kick and toms), and set the touch sensitivity of the sampler to zero. This way, every kick will trigger an equally loud sample, which opens the gate the same way every time. Such clear trigger signals can also be made by recording the drums with additional contact microphones or drum triggers, attached to the rims of the drums. And many DAWs have an audio-to-MIDI function that uses transients to generate MIDI notes with velocity values that can be manipulated. The major advantage of this approach is that you can easily correct double triggers or missed drum hits by hand.

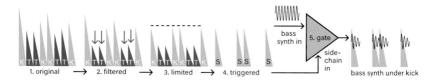

Figure 9.21 Creating a stable trigger signal to precisely place a bass synth under the recording of a kick. First you EQ the kick to attenuate the spill of the toms as much as possible, and then you use a limiter to make the kicks equally loud, so they will form a reliable source for triggering a sample on every kick. Since this sample contains no spill, the gate won't get 'confused,' and it will react consistently on every kick hit.

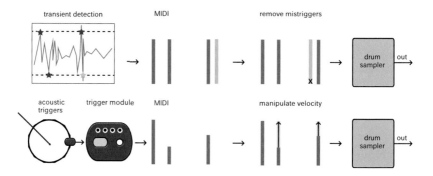

Figure 9.22 By using transient detection in your DAW or a trigger plugin with a MIDI output (top), or acoustic triggers and a trigger module (bottom), you can generate MIDI from your acoustic drum sounds. This makes the manipulation of the trigger points a lot easier.

Touch Sensitivity

It's a different story if you want touch sensitivity, for example if you intend to replace or supplement the sound of a particular drum without changing the dynamics of the recording. Good source material is very important in this case, because you don't want to go overboard in terms of filter manipulation (as this changes the dynamics), yet you should still be able to distinguish a soft snare drum from a loud hi-hat. This is often a problem, since the weak recordings that you want to supplement with samples usually have a lot of spill. This means that sometimes there's no other option than to align the samples manually with the waveforms, and to scale them in terms of dynamics.

Making Samples

Of course, touch sensitivity is only interesting if the things you trigger actually do something with those level variations. For example, you can use one instrument to trigger the compression on another instrument through an external sidechain. In this case, the louder the first part plays, the more the second one is compressed. Or you can make the drums trigger the gate on the reverb, and the louder the drummer plays, the longer the gate remains open. If you want the triggered samples to have convincing dynamics, it's important that you use samples that were recorded at different loudness levels (or create them yourself).

A softly played snare drum doesn't sound the same as a snare that was played loudly and then turned down. For this reason, it's a good idea and a small effort to ask the drummer at the end of a recording session to play all the drums individually at a few different loudness levels. Sometimes these samples can already be used in the same project to drive reverbs or trigger extra ambience. In any case, they will be a nice addition to your library, so you won't have to use the same drum samples as the rest of the world for your next project.

9.5.2 Triggering in a Mix

Building Drum Sounds

The perfect drum should work well in the music on various levels: its center of gravity and resonance must match the style, the part and the tempo, there should be enough attack for it to cut through the mix, and if it also sounds unique, that's a nice bonus. However, it's not always possible to tick all these boxes with a single drum sound, whether it's a recording or a sample. In these cases, you can trigger additional samples (or a drum synth) that you combine with the original. A sample with a clear, sharp attack can help a round, warm kick to cut through the mix, while retaining its original character. It's very important to keep an eye on how the different drum sounds add up, because if you're not careful, they might end up attenuating each other instead of amplifying, due to phase interaction.

It helps to strongly filter the samples that you use as supplements, so they will only add sound in the ranges where you need it. Polarity switches and all-pass filters can help to get the lows in phase, and pitch shifting the samples can make them match well in terms of tone. A trick that won't cause any phase problems is triggering a gate on a noise generator. The resulting envelope of the pulse can be perfectly controlled with the timing of the gate. And with the help of some EQ and reverb, you can easily turn an ordinary snare drum into a cannon.

Adding Space

Sometimes your drums lack space and size, but adding reverb turns your mix into an undefined mush. In these cases, triggering can help to use the space more efficiently. Using an extreme gate on an acoustic snare drum usually doesn't sound very great, but if you only use the gate on the send to the reverb, you'll make sure that only the snare—and not the spill—ends up in the reverb. This way you can make the snare drum appear to be louder in the recording room than it actually was, without manipulating the dry signal. On drums, a distant room microphone usually sounds better than artificial reverb, but the sound it picks up tends to be dominated by the cymbals, so it instantly fills up your mix. You can avoid this by using the close mic of the snare to trigger a gate on this ambience. And by stretching the gate's attack, you can even remove the first part of the room's reflections, which subjectively makes the space seem a bit larger and more diffuse. This effect sounds very different from delaying the room microphones to achieve pre-delay, because in the latter case, you'll still hear the early reflections, only delayed. Using a gate doesn't mean that you need to completely close the gate between the snare hits. An attenuation of 6 dB already helps a lot to keep the sound focused, while the snare does sound a lot more spacious. If the room doesn't sound well, if the drummer's playing is out of balance, or if you don't have separate tracks of everything (drum loops can be tricky when it comes to this), you'll reach the limits of the recorded material. You can then try to make each drum hit trigger a room-miked sample of a similar drum in a (larger) space. This can be a pretty laborious process, if you want to do it convincingly (it sometimes helps to compress the triggered ambience together with the drums), but the result can sound very impressive, yet still defined (see Figure 9.23).

Movement in the Mix

When good musicians play together, they constantly anticipate each other. This almost automatically results in a mix that's full of cohesion and interaction. The more 'production' you add in the form of overdubs, extreme effects or programmed parts, the less of this kind of natural interaction you'll get for free. Fortunately, there are techniques to generate a bit of interaction between different parts. The most well-known of these is sidechain compression, which is when one instrument pushes another away. But the opposite can be just as useful, for instance when a rhythm triggers an expander on the bass, which then ends up giving the rhythm an extra boost. This way, you can make a sustained bass sound as if it was played with a similar rhythm to the drums. In an extreme form, with a gate instead of an expander, you can use a trigger signal to turn a particular part on and off. The famous 'trance gate effect,' which is when a fast rhythm part makes a pad sound stutter, can be created this way.

Interaction doesn't necessarily have to take place between existing parts. As a producer/mixer, it sometimes helps to think more like a

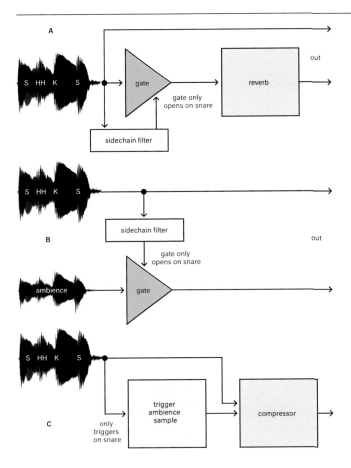

Figure 9.23 Three ways to add space to a snare drum recording without emphasizing the cymbals too much. A: Gating the input of the reverb, so only the snare goes into the reverb and the kick and hi-hat don't. B: Putting a gate on an ambience microphone, and only opening it when the snare plays. C: Using the original snare to trigger an ambience sample of a separately played snare.

musician, and start interacting with an instrument or effect yourself. By triggering EQs, compressors and expanders manually, you can automatically make parts less flat, as if the performances were much more high-octane than they actually were. A boost in the high mids on all the guitars can be too much if you apply it to the entire chorus, as this would leave no room for the vocals. But if you only trigger it on the first attack and maybe again halfway, you can create the impression that the guitars play a lot louder in the chorus than they actually do. A simple patch for this is shown in Figure 9.24.

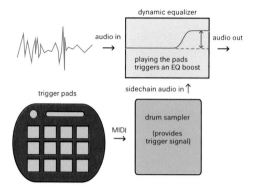

Figure 9.24 Instead of using another instrument to drive the sidechain of a dynamics processor, you can also play a sampler to easily trigger gates, expanders or dynamic equalizers yourself during mixing. With the construction above, you can give a certain part a bit more bite, but only at select moments. In this case, the drum sample only serves as the command signal for the dynamic equalizer.

9.6 Spectral Editing: Unreal Reality

Acoustic sound consists of an interplay of resonances, each with its own pitch, intensity and progression in time. If you can manipulate the individual resonances of a sound, you can dramatically influence its character. For instance, by giving a trumpet a different overtone structure, or by suppressing the soundbox resonances of an acoustic guitar, you can create instruments that sound strangely familiar, but not quite. Spectral processing makes this kind of magic a reality, but you'll still have to find useful purposes for it.

9.6.1 How it Works

You can visualize sound in two ways: in the time domain or in the frequency domain. I'm sure you're familiar with the time domain from working with your DAW: an audio track displays the amplitude of a sound wave, which moves from left to right in time. The frequency domain is a bit more abstract, as you don't see the same sound wave at a single moment in time, but in its entirety. And not as a complex wave, but as a long list containing the frequencies and amplitudes of all the sine and cosine components that are present in the sound. According to Fourier's theorem, any sound can be described as a combination of sine and cosine waves. Converting a sound to the frequency domain is done with a (fast) Fourier transform. And after you've done your magic with it, you can use the inverse Fourier transform to turn the list of frequency components into sound again. This transform/inverse transform is lossless if you don't make any changes, so the sound will come out the same as it went in.

All spectral effects processors use a Fourier transform to gain access to the individual frequency components of a signal. But what they do with this information differs widely. Sometimes the manipulations have more in common with the formulas in a spreadsheet program than with traditional effects. For instance, you can shift or exchange frequency components. You can erase, copy, stretch or delay them, or manipulate their intensity (see Figure 9.25). Time and pitch are very elastic concepts in the frequency domain. In order to visualize the data—and the manipulations you make with them—a graphical representation (a spectrogram) is often used. Sometimes you can directly make edits inside this graph, such as selecting a specific frequency component or drawing a filter response.

It's easy as pie to create completely unrecognizable sounds with only a few simple edits, but their usefulness is limited. It gets more interesting when the analysis and processing of the data are dependent less on random craziness and more on recognizing and manipulating specific aspects of the sound. For instance, the detection (and therefore isolation) of individual notes (including all their overtones) in a harmony. Or the detection of a spectral envelope, which is the filter response of

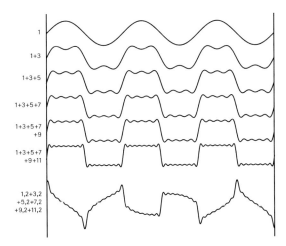

Figure 9.25 Spectral processing can be used to easily manipulate overtones. In this example, a square wave is built up of partials. Eventually, each of these partials is moved up by the same number of hertz (frequency shifting), which is not the same as shifting the entire square wave up (pitch shifting). The ratio between the partials changes, resulting in a completely new waveform, as you can see when you compare the bottom two waveforms.

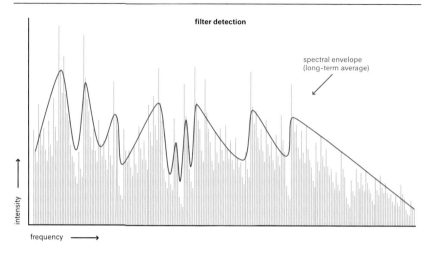

Figure 9.26 A spectral envelope can be analyzed and then used as a filter, for example to make a flute sound as if it was played through the soundbox of a violin.

the soundbox that shapes the source sound. Examples of this are the resonances of your voice box, or those of a snare drum's shell. If you can distill this kind of filtering from the source material, you can also apply it to other sounds. Or you first remove the resonances of the original soundbox, and then manipulate the filter coefficients (if you realize during mixing that you should have used a deeper snare drum shell) and apply these to the source material (see Figure 9.26).

9.6.2 Spectral Processing in a Mix

Laser EQ

A spectral processor can be seen as a very finely divided crossover filter that can give each frequency band its own manipulation. I will describe some of the exotic possibilities this has to offer, but first the most basic manipulation: amplifying or attenuating a particular frequency band. Of course, you can also do this with a normal equalizer, but removing one specific tone plus all its overtones can be done faster and more accurately with a spectral EQ. This is in part thanks to the visual feedback, but mainly to the precise selection possibilities and steepness of the filter slopes. On top of this, any filter curve you can think of is easy to make. You're not tied to standard shapes like shelf or bell, but you can freely draw all kinds of asymmetrical shapes. Because of this, you can see a spectral processor as a graphic equalizer with 512 to 4,096 bands. This allows for very distinct-sounding EQ treatments.

Repair

Most spectrum editors were originally intended as a repair tool. They can be used to fix all kinds of disturbances in your recordings. As long as these are loud enough to attract attention, they can be isolated by drawing a selection frame around them. The exact frequencies of the disturbance can then be attenuated or replaced by an adjacent sound fragment. Some editors have smart selection tools that make this process even faster and more focused. For example, they let you select a tone and all its overtones in a harmonic series with one click, or all frequency components that last longer than a specified time. These smart tools make it tempting to use these editors as music manipulators as well, instead of only removing coughs with them. Stretching one note of a chord in time, amplifying or shifting specific overtones, seamlessly replacing a piano's hammer with a tone so it becomes a kind of organ; anything you can think of can be done with spectrum processing.

The graphic way of working has inspired some manufacturers to add an import option for graphic files to their spectral (synthesis) software. This usually won't lead to sound that's immediately usable as music, but images with regular patterns can sometimes produce surprising rhythms or overtone structures.

Dynamic Wizardry

It gets even more exciting when you start amplifying or attenuating the frequency components signal-dependently. For example, a spectral gate can remove all the frequency components from the signal below a user-defined threshold. This is the system that's used by a lot of noise reduction plugins, which mainly differ in how intelligently they adjust the threshold of the various frequency bands to the (often variable) source material. Besides as a tool for attenuating noise, spectral gates are also very interesting as an effect. For example, you can use them to trim a sound down, so only its tonal essence remains. This can make the sound more condensed, and therefore more focused.

The reverse manipulation—attenuating all frequency components that exceed the threshold—is also possible. The spectral compressor that you create this way can be seen as a multiband compressor with an infinite number of bands. The great thing about this is that you can deduce the threshold level of the individual bands from the source material. If you

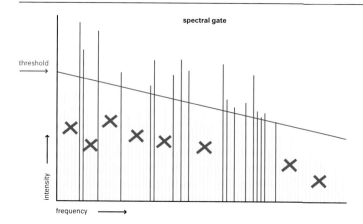

Figure 9.27 Noise reduction software removes frequency components that fall below the threshold, but a spectral gate like this can also be used as a dynamic filter effect.

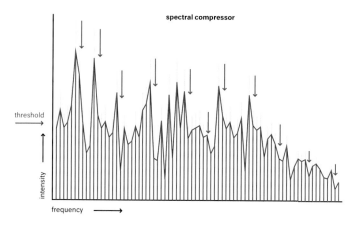

Figure 9.28 A spectral compressor is a multiband compressor that can guard the frequency balance of the source sound very accurately.

like the tonal balance of a certain section of the material (for example, a vocal line without hard S sounds), you can use this section as a benchmark. As soon as there's a passage with a lot of sibilance, the compressor will only work in those ranges. Pro Audio DSP's Dynamic Spectrum Mapper is a good example of such a compressor.

A Different Kind of Echo

So far, all frequency components have neatly stayed in their place on the timeline, while only their amplitudes have been adjusted. But it's also

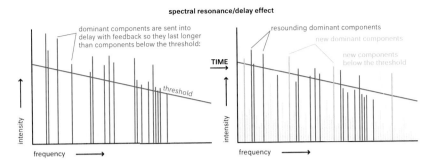

Figure 9.29 By sending frequency components that exceed a threshold level into a delay with feedback, you create a resonance effect that only makes the dominant frequencies resound longer.

possible to move individual components in time, or send them into a delay with feedback. This way, you can produce sounds with the timbre of the source sound (so they will still be vaguely recognizable), but with a completely different progression in time. Strange envelopes, resonances, time stretch-like effects and even reverb can be created this way. Spectral delay is a particularly great effect for sound design, because it's a perfect way to take a sound that evokes an overly conscious association, and turn it into something subconscious. You'll still hear the original character, but you can't actively identify the source sound anymore.

Up- and Remixing

Besides as an effect, spectral processing can also be used to isolate sound elements in a stereo mix and redistribute them across more than two speakers (upmixing), or to isolate only the vocals so you can use them in a remix. This is not so easy, and you can see it as an attempt to undo a finished mix. Still, with modern software you can achieve pretty useful results. Especially by analyzing the direction (based on phase and intensity differences) of all the frequency components in a stereo mix, you can find out which components belong together. In simple terms: if two frequencies always sound at the same time and come from the same direction, they are probably produced by the same instrument. By detecting melody lines and recognizing which transients belong to which notes, the detection can be further refined.

Smooth Transition

The frequency spectrum of a sound can be seen as the sheet music for all the frequency components that make up the sound. If you had a bank with oscillators, one for each frequency, this spectrum would precisely indicate which oscillator should be activated when and at what intensity in order to reconstruct the sound. That's the exact same principle a vocoder uses to apply the character of one sound (often a voice) to another. The vocoder measures the distribution of the source sound in a number of frequency bands, and then applies this distribution to the target sound. A spectral vocoder works the same, but with a much higher degree of precision than the classic vocoder, simply because the number of frequency bands for the analysis and reconstruction is much larger.

This higher resolution is not the only advantage of spectral vocoding. Much more complex manipulations are possible than with a classic vocoder, such as seamlessly blending one sound into another. This principle is called morphing, and in the simplest form it involves two frequency spectra that are merged into each other by filling in the intermediate spectra (through interpolation). This way, you can create very interesting effects. A rhythm guitar could start a song, but then slowly turn into a drum kit that plays the same rhythm. The provocative lyric you always wanted to incorporate into a track can almost unnoticeably be 'uttered' by the rest of the instruments. Seamlessly blending multiple vocal characters together to create a schizophrenic impression (or to symbolize the ultimate togetherness, depending on how you look at it) is a possibility as well. And even making a transition from one complete track into another can yield interesting results. At least, it will be something different than sweeping your filter down while you mix in the next beat. Not all morph effects use the same technology, and a simple interpolation of two spectra won't always lead to a convincing intermediate form. That's why smart algorithms use psycho-acoustic principles to determine which sound aspect should be dominant in the transition from one sound to another.

Chapter 10

Automation
Performing the Mix

Music changes constantly, which might be the main reason why you want to listen to it. Therefore, there's no such thing as finding a position for your faders, pan pots, equalizers, compressors and effects that's right for an entire song. You could even say that a static mix is unmusical by its very definition. That's why this chapter is about following the music with your mix. This might seem like a lot of work at first, but once you get the hang of it, it will feel liberating and work faster to constantly adapt your mix to the music. For example, you can try to make the vocals blend with the music by running them through three compressors and a de-esser, but after an hour of looking for the perfect setting, you'll often find that your equipment doesn't understand the musical context at all.

When a singer lets out a high-pitched wail, maybe it's supposed to sound loud and shrill. This is something a compressor will never understand, as it only looks at the signal level. But if you close your eyes and use fader automation to place the vocals in the mix, you can make the entire dramatic arc of the vocal performance work in the context of the mix. Usually, this will only take three rehearsals (plus maybe redoing a few short sections). This technique is called fader riding, and besides saving time, it can make the music more dynamic and lively than a machine-based approach. If you can immediately tackle a problem with a fader movement when you first hear it, you'll be less tempted to manipulate your mix in all kinds of ways that will only end up making it sound flatter. This chapter is about making your mix playable, so the—inherently musical—manual solution will also be the easiest solution.

10.1 Faders for Each Function

When I first started to play the drums, my main goal—besides playing as fast as I could—was to expand my drum kit so it would occupy as much floor space as possible. Most mixing engineers will probably recognize this insatiable hunger for new possibilities. The drawback of all those extra cymbals and drums was that I was spoiled for choice. At some point I noticed that a well-placed rhythm using only a few elements could be much more powerful than something spread out across the entire arsenal. There was a lack of consistency (and maybe of vision as well) when things became too complex. There's something very organized about only using three drums: the part you play becomes much more important than the drums you play it on. I feel that it works the same way in the studio. When the possibilities are endless (whether these are soft- or hardware), you tend to forget what the music actually needs.

In the studio I try to get back to the essence, just like I did with my drum kit. Not necessarily by using less gear, but by organizing the material and making the equipment easily playable. This means that the setup should invite you to grab the faders and play with them like you would play an instrument. If you suddenly realize the drums should be bigger in the bridge, you want to try this right away. If it works, it's great. And if it doesn't, it won't take much time and you can quickly move on to the next idea.

One of the simplest things you can do is grouping tracks so you can manipulate all of them with a single adjustment. The best-known example is to send all drums or background vocals to one group so you can automate them as a whole. But sometimes an instrument (group)-based division isn't the most musical solution. Maybe you want to boost all the rhythmic elements in your song with one fader, or fade out all the low-frequency instruments. Or leave the essential parts alone, while you slowly add all the overdubs, harmonies and effects.

VCA faders are fun tools that make such a musical function-based division possible. With a single movement, a VCA fader controls the fader settings of all the channels you assign to it. This is not the same as creating groups that you actually send the audio through. When I make audio groups, I do this based on the manipulations I want to perform on the group as a whole, while my VCA groups are based on the movements I want to make in the mix (see Figure 10.1). This process requires some experimenting, as you first need to find suitable movements in sound for the particular music you're working on.

Once I've discovered a couple of these 'flavor enhancers' after some research, I make sure they're organized in the project, so I can control them with a single fader. This way, you're basically building a customized, easily playable controller for each mix. If it's inviting you to use it, you're bound to deploy it as a musical tool. But if it's too complicated, you usually won't get around to it.

10.2 Parallel Constructions

As well as level automation, you can also use parallel-connected effects processors to create contrasts that help propel the music. Distortion to

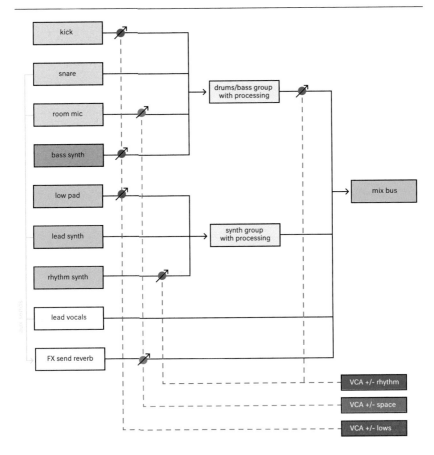

Figure 10.1 VCA groups help to make your mix manageable: with only a few faders you can make huge differences, for example by fading out all spaciousness or boosting all rhythm parts at once.

add intensity, delays to introduce a rhythm, an exciter to make things sound closer, extreme compression to blow something up, a reverb with feedback to make accents longer, a flanger for stereo width, and so on. Because the effects are connected in parallel, you can easily automate the ratio between the original and the effect with the fader of the effects return.

You can take it one step further by setting the effects send to pre-fade and then automating three things: the level of the original, the level of the effects send and the level of the effects return. For example, you can have a lead vocal start without compression. Then you add more and more parallel compression, and turn the send to this compressor up so far that it starts to distort. Finally, you slowly fade out the original signal and turn up the effects return. This will make the distortion dominant, but without changing the overall volume of the vocals in the mix. Signal-dependent effects like distortion or compression will become even more expressive with an approach like this (see Figure 10.2). Another advantage of automating both the send and return level of an effect is that you can put reverb on a single note (and not on the rest of the part) by only opening the send during that note. Conversely, if you want to insert an abrupt silence, the easiest way to this is by turning down the reverb's return. If you only closed the send, the tail of the reverb would still be audible during the break.

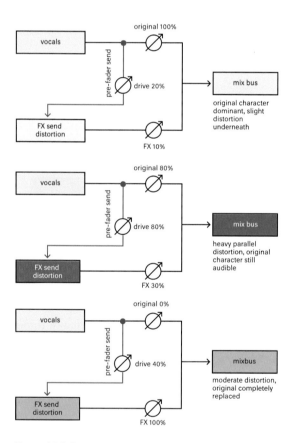

Figure 10.2 By routing send effects pre-fader and then automating the original sound, the effects send and effects return, you can easily add contrast to your mix.

Controllers

In theory, you don't need a controller to work with your DAW. It doesn't add features that you can't control with your mouse and keyboard. And for the transport buttons you don't really need it either. Still, a controller with motorized faders is essential for one reason: it's the only device that lets you automate your levels with enough speed and precision to actually want to do this in practice. If you've ever tried to do vocal rides with your mouse, you'll understand this immediately: sometimes one decibel makes all the difference, and other times you want to boost a word by six decibels and return to the original level half a second later.

A fader is the only tool that allows you to keep an eye on this sense of scale; with a mouse, you'll quickly be swerving all over the place. On top of this, you can control two to three faders at once. And no, it's not the same thing to have a controller with virtual faders running on an iPad (or another multi-touch screen). Due to the irregular friction between the screen and your fingers, your connection with the fader won't feel nearly as direct. The only alternative that comes close to this feeling is an analog potentiometer. Endless encoders are less suitable, as they generally skip values if you turn them too fast. What you want is a real variable resistor with a fixed starting and end point. Ideally, this resistor will also be motorized like a fader, so it's always in the right initial position. The most frequently used component on my own controller is a motorized potentiometer that always follows the most recently manipulated parameter. Controllers that can map other parameters to their faders than the channel levels are also perfect for this. This way, you can automate aux-sends and make monitor mixes in no time.

Diffusion

Every time you use an effect in parallel with the original signal, there's a danger of unwanted phase interaction between the two. Effects that are based on relatively long delays (like reverb or echo) won't cause this problem, as these effects are sufficiently disconnected from the direct sound. With effects like chorus or phasing this interaction does occur, but it's actually desirable for the change in timbre it causes. It only starts to become problematic when you use parallel distortion or filtering, or when you try to create a doubling effect with very short delays or mild

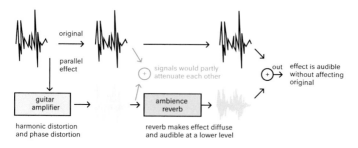

Figure 10.3 A short reverb can 'disconnect' two signals that would otherwise partly attenuate each other. The result is a small change of timbre, but it does make the two signals add up better.

pitch shifting. In these cases, the original and manipulated sound are correlated (see Chapter 7). Their interaction can lead to attenuation and amplification of certain frequency ranges.

To avoid this, you can try to align the signals as best you can in time (with delay) and in phase (with all-pass filters), but sometimes you simply can't make them add up nicely, no matter what you do. The opposite approach can work well in these cases: instead of trying to get the original and the effect perfectly aligned and in phase, you disconnect, or 'decorrelate' them. One way to achieve this is by processing the effect with a very short ambience reverb. This will alter the effect in terms of phase and time, mixing it up in a way. Of course, this also adds some space to the sound, but there are reverbs that add relatively little colorization and decay, while they still make the signal more diffuse (see Figure 10.3).

Sculpting

To make a parallel effect blend well with the original sound, it's important that it mainly fills the areas where there's still a shortage. Sometimes the loud, long notes are still a bit bare, and at other times the soft, low notes lack some texture. In one situation you want to thicken the sound with an overlapping effect, and in another you're trying to fill the gaps between the notes with a separate reverb. Equalizers, (multiband) compressors, delays, gates, expanders and envelope shapers can all come in handy if you want to maneuver an effect to a particular location in frequency and time. Just take a look at the example in Figure 10.4, which shows how you can feed and manipulate a reverb in such a way that it exactly fills the space you had in mind, and not an inch more.

Some delay effects come equipped with gating and ducking options. They look at the input signal and then turn down the delay, either when

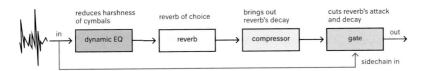

Figure 10.4 **By using processors before and after the reverb send, you can precisely shape the reverb in terms of timbre, dynamics and timing.**

there's no signal coming in (gate), or when there is (ducker). This way, you can easily make sure that there are never any loose echoes conspicuously bouncing around. Or you can prevent the echoes from getting in the way of the original signal, when you only want them to fill the gaps in the performance.

10.3 Change and Conquer

The worst thing that can happen in music (or in any other story you're trying to tell) is that it gets boring, and the listener tunes out. This best way to avoid this is by regularly introducing new events, which can give the story more depth or a new turn. The introduction of new events is primarily the job of the composer, songwriter or producer, but these events (transitions) will be much more powerful if they're supported in the mix. This support doesn't even have to be exotic to be effective. Simply automating the cutoff frequency of a high- or low-pass filter can really emphasize a dynamic movement, for example. Boosting the introduction of a new part and then turning it down will make the transition much stronger. And you shouldn't underestimate the effect of panning: going from narrow to wide (or vice versa) hugely changes the contrast. These kinds of movements will be even more effective when they all work together. If you're a bit of a programmer, you could even try to build a 'meta controller' for this purpose: a patch (a controller setting) that will pan, amplify and filter the sound with a single fader movement.

Some plugins have the option of morphing between two presets, which means that you can use a slider to create a transition from one setting to another, or choose any position in between. This only works with effects that were built in such a way that a parameter change never produces a click in the sound. Scene automation in digital mixing consoles also gives fantastic results: with one push of a button, you can move from one mix to another. The only problem is that the transition is instant, so you can't

Automation

There are a couple of frequently used approaches to parameter automation. These approaches mainly differ in what happens after you let go of the parameter. With touch automation, the parameter then jumps to the value it had before you started automating it. This is typically a mode I use when I've already finished the automation of a song, but I still want to boost a few elements a bit more. At these moments I push the fader up a bit, and as soon as I let it go, it jumps back to the original automation curve. This is different in latch mode, in which the last value of an automation take is maintained until you stop the transport or take the channel out of write mode. And finally, when you're completely satisfied with the course of an automation curve, but you still want to give it a certain offset, the trim mode will come in handy. In this mode the original curve is maintained, but shifted up or down depending on the trim automation you're recording.

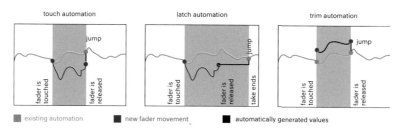

Figure 10.5 Frequently used automation modes. The exact implementation of the automation varies per system, for instance the way the values for empty sections are determined.

make long morphs. This is a personal wish of mine for future DAWs: a customized fade curve to move from one mix to another, so a kind of VCA fader for all mixer parameters, including plugins. I'm afraid that's wishful thinking though, given the fact that only a few plugins offer the possibility of creating smooth parameter changes. Some settings switch in steps, and adjusting buffer sizes or delays will result in dropouts or pitch shifts, for example. Therefore, the next best thing is to use automation for every effect that supports smooth changes.

Some DAWs let you change the settings of a plugin in preview mode, and then store these changed settings as automation with a single click.

Daniel Lanois

An inspiring example when it comes to dynamic mixing (and doing more with less) is Daniel Lanois. This producer/mixer/musician (U2, Brian Eno, Black Dub) has put a couple of nice videos online that showcase his improvisational way of mixing. A few fairly simple devices (an old and quite primitive Lexicon delay, an Eventide H3000 and a Korg delay) are the tried and trusted building blocks of his mixing setup. Instead of looking for the ideal setting for a particular mix, he knows these devices so well that he continually adjusts the settings as the mix develops. Every performance he does this way is unique, and if he wants to use any of it, he'll have to record it. The movements he makes on these devices can't be recorded and edited as automation. This might seem like a very primitive and restrictive way of working—because it is—but it does lead to extremely lively, dynamic and unique mixes.

This is very similar to scene automation, but per plugin and with the option of editing the course of the automation afterwards. Especially if you want to morph from one EQ setting to another, this is a great approach. It's a widely used postproduction method in film and television, used for giving sound the right perspective when the camera position changes.

10.4 Expressivity

When I automate effects, I often think of the expression of a good singer: all the character you need is already in that sound. In a production, you usually only need to balance it with the accompanying instruments, give it a spatial position and maybe bring out some underlying details (unless you want to give it a completely different character for productional reasons). A voice can be seen as a very complex effects device, with variable filtering, dynamics, pitch shifting and harmonic distortion on board. On top of this, the sound of this voice is directly controlled by the brain of a musician (who is also telling a story that listeners understand, and that they can even imitate with their own voice). Together, this produces a complex sound that can easily keep listeners interested. But how do you achieve similar expressiveness with a sound that has less character in itself? How can you make a flat-sounding organ sing?

Signal Dependency

Part of the expressivity of a voice is the result of the sound changing dramatically when the vocalist sings soft or loud. Not only does the loudness increase, but the harmonic structure, the relationship between vowels, consonants and other noises, and the filter character of the oral cavity change as well. Some effects (especially distortion) also change in timbre when you increase the input level. This principle can come in handy if you want to make other effects more expressive. For example, reverb is typically an effect that doesn't change in timbre when you drive it harder. If you add some distortion to the signal before you run it through the reverb, you can make the character of the reverb change along with the input level. Figure 10.6 shows a slightly more complicated circuit, in which the source signal not only affects the harmonic structure, but also the dynamics and the frequency spectrum of the signal that goes into the reverb. Many effect parameters lend themselves to being controlled with an envelope follower: filter frequency and resonance (if you do this subtly, you can also create effects that are a bit less conspicuous than the well-known auto-wah), reverberation and delay time, feedback, modulation depth, the amount of pitch shift in a delay's feedback loop, and so on. Once you get the hang of it, you'll immediately understand why modular synths are so appealing: virtually all parameters can be controlled with any source you want.

Complex

By nature, people are focused on perceiving connections between different elements. Regular patterns and correlated movements easily stand out. The more predictable the patterns, the quicker they draw your attention—and the sooner you'll get bored. I'm not arguing that regularity

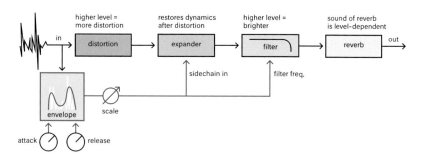

Figure 10.6 A way to make the sound of a reverb effect respond to the input level.

should be removed from music altogether, because 'recognizable patterns of sound' is actually not a bad definition for the concept of music in general. But if you want to use effects to add a deeper layer to the music or to place it in a particular context, you'll often look for sounds that are not so predictable. In other words: you can paint sailing ships on a perfectly repeating pattern of waves, but your composition will really come to life if all the waves are unique, refracting the light in different ways. It will almost make you feel the wind in your face, with a little imagination.

Adding ambient noise to music is one of the more popular ways to give it an interesting twist, but if you're not careful, it easily becomes a gimmick because it can also evoke very concrete associations. You don't want to immediately feel like you're standing in a mall or walking alongside the water. Therefore, many producers who want to add an 'atmospheric layer' set out to find sounds that lend themselves to this purpose: abstract sounds that you can attribute your own meaning to, that don't have a lot of regularity or familiarity. For instance, a pad sound with a strong sonic development, or an improvisation on an instrument with different feedback-heavy delays. A reverberating piano with so much compression that it produces a cloud of modulating resonance, or noise that goes through a chain of moving filter banks and reverb; anything you can come up with to build an underlying atmosphere that glues the other sounds together.

Reverse delays (which are easy to make with some granular delays that cut the sound into little pieces they then reverse) and reverse reverbs (just record some reverb on a separate track and reverse it) can also work well to make the meaning of a sound less concrete and stretch it into an underlying atmosphere. A combination of effects usually works the best, because this will make it easy to add enough movement and unpredictability to the sound. Plus I often equalize the final result, so it will exactly fit around the rest of the instruments. It's as if you're adjusting the acoustics of an imaginary space to the timbre of the music.

To me, creating atmospheres with electronic effects is one of the most fun things you can do in the studio, because you can make it as crazy as you want. Even in acoustic music these kinds of underlying layers have a place, though often only as a kind of background noise. In sample-heavy genres like hip-hop it's completely normal to hear different backgrounds being turned on and off over the course of a song. Sometimes this will sound like a collage of different atmospheres, and the hard transitions between these atmospheres start to contribute to the feel of the music in their own way.

Of course, you can also apply this principle to your own atmospheric sounds: if you suddenly cut them or turn them on at strategic moments,

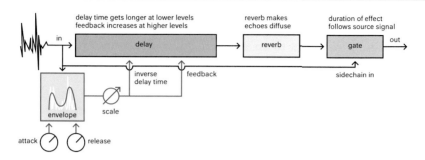

Figure 10.7 A delay effect with a signal-dependent response, which precisely follows the dynamic course of the input signal.

this can sometimes make the music even more expressive. It creates contrast, and it will make your transitions harder and more powerful. Figure 10.7 shows a simple patch to make a rhythmically modulating atmospheric effect follow a sound.

Chapter 11

Advanced Techniques
Exploring the Boundaries

This chapter is about the boundaries of your mix. What happens in the lowest and highest frequency ranges, for example? How can you use these frequencies to manipulate the size of your mix? And how can you control annoying frequency peaks that draw your attention away from the music? What determines the size of the stereo image, and how can you make sure it's both focused and wide? And to what extent does this change when your listeners use headphones?

11.1 The Floor: Keeping Low Frequencies in Check

There is something elusive about low frequencies. If they behave well, they can make huge crowds dance, but if they don't, they will flood your mix in no time. It can be pretty difficult to control them, not least because the low end sounds different everywhere, due to the influence of acoustics and speakers. If only you could ignore the bass region altogether, then at least you would be sure your mix sounds roughly the same everywhere. But even in a mix in which none of the instruments play low notes, the low end is important. If you remove the frequencies below the lowest fundamental, the rhythm loses its power, the sense of spaciousness and size disappears, and—though you might expect otherwise—the midrange often ends up sounding restricted as well. So there is no getting around it: you will need to look for ways to control the low end.

In Balance

Some of your listeners have their subwoofer placed in the corner of the room, resulting in overemphasized lows. Others can't even perceive certain frequencies in the middle of their perfectly square house, due to its acoustics. So it's quite a challenge to satisfy both of these extremes, and you'll never fully succeed at this. What you can do is make sure that the lows in your mix are so well balanced that they won't ex- or implode in

---

Am I Hearing This Right?

If you often find yourself using narrow filters to dip the same resonance, there's a significant risk that you're correcting your own acoustics and not the source sound. This is because these kinds of extreme problems in your source material are pretty rare. If you're not sure, a minor correction with a wide filter is a safer choice to tame the lows a bit, as this will result in less phase distortion than when you use narrow filters. Checking your mix in another room or on headphones can also help to get a better idea of the problem.

---

these situations. But what does it mean to have a well-balanced low end? One mix might need more lows than the other, depending on the genre and arrangement. Therefore, there's no absolute measure for the amount of low end, and a spectrum analyzer won't help you either. The key is in the relative proportioning of the lows: what's on the other end of the spectrum to keep them in balance? These proportions are musical rather than technical. For example, turning up the vocals can also be a good way to make the lows less dominant. If your own acoustics aren't reliable enough to judge the low end, it can help to assess the musical proportions on speakers that don't reproduce low frequencies. If you create a good balance this way and refrain from using extreme EQ in the low end while doing so, chances are that your mix will translate quite well to other situations.

Spreading Your Chances

Acoustics are the largest variable when it comes to low-range reproduction, but also the least predictable one. In the car, one frequency is emphasized, at home another. This isn't necessarily the end of the world, except when your mix is heavily based on a single frequency. For instance, if you use your synthesizer to program a kick with one particular fundamental, it can completely disappear in one space and strongly resonate in the other. But if you connect an envelope generator to the pitch of the oscillator you're using, you can create a kick that's spread across more frequencies. For example, by bending the pitch slightly downward after each hit, the kick will no longer have a single fundamental, but an entire range of tones (see Figure 11.1). If these pitch bends are too extreme, the result won't always be musically desirable (although they will hardly stand out as something tonal in the sub-bass range—many acoustic drums exhibit a similar fall in pitch when their notes decay). Another way to

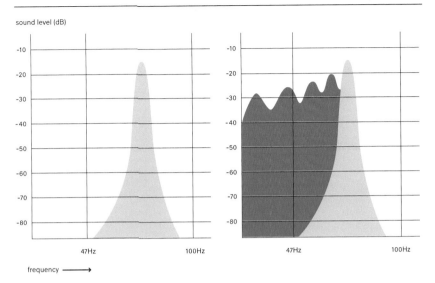

Figure 11.1 If you use a dynamic pitch shifter to add a slight pitch progression to the blue 'sine kick,' the newly generated red frequencies will make sure there's always something to be heard, regardless of the acoustics in different listening rooms.

enhance the sound is by combining multiple kick samples. This wider distribution of low frequencies is a way of spreading your chances: instead of relying too much on one particular frequency, you use an entire range. The result is that there will be something useful in the sound for nearly all acoustic situations: if one frequency is attenuated, there's always another one to take over.

Overtones

But what do you do if stacking sounds is not an option, for example if you recorded a double bass that should still sound like a double bass? In these cases, you can try to enhance the harmonics of the sound by emphasizing its overtones. Nearly all acoustic tonal instruments—including the human voice—have a series of overtones at fixed intervals above the fundamental (see Figure 11.2). You're so used to the sound of these intervals that if you only hear a part of a series, your brain automatically fills in the rest. A result of this is that instruments with relatively strong overtones are much easier to perceive in a mix than a sine bass, for example. However, you can help this sine wave a bit by running it through a tube amplifier or tape recorder, thus adding some harmonic distortion (which also consists of overtones). This will make the bass more easily audible, even at low monitoring levels. The result is that

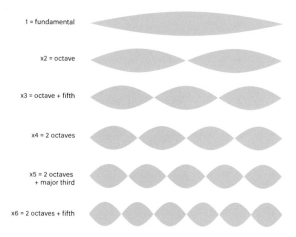

1 = fundamental

x2 = octave

x3 = octave + fifth

x4 = 2 octaves

x5 = 2 octaves
+ major third

x6 = 2 octaves + fifth

Figure 11.2 A natural series of overtones, as produced by a string instrument or vocal cord. If you heard this series without the fundamental, you would automatically add it in your head.

you can turn it down in the mix, leaving you room for other sounds. Plus there's another advantage: overtones help to make the low end more audible on small speakers. They suggest the presence of a fundamental that's too low to be reproduced by the small speaker.

Processors like MaxxBass by Waves are also based on this principle, but only add overtones to the deepest frequencies. Unlike an overdriven tube amplifier that distorts the entire spectrum, the sound remains unaffected in the mid and high ranges, while the low end does become harmonically richer. A similar effect can be created by running the output of a tube amplifier through a low-pass filter, and combining the result with the original signal. This is a very useful method to keep the deepest frequencies sufficiently 'connected' with the rest of the frequency spectrum, plus it's a way to make sure that when you listen on headphones or on a television, you'll still feel there's something impressive going on in the low end.

But there are drawbacks as well: the overtones can take up too much space in the lower midrange, which is usually already densely packed, since most instruments have their fundamentals in that area. On top of this, adding harmonic distortion to generate overtones reduces the dynamics of the signal. Technically, the process is nothing more than (level-dependent) soft clipping, and you need a lot of it to also affect the soft notes. Adding the distortion parallel to the original signal can help to retain some of the dynamics. Especially if you first even out the parallel signal with a compressor before you distort it, it's easier to give each note the same amount of distortion (see section 9.1 for more on this topic).

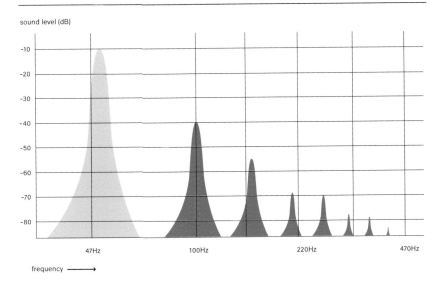

sound level (dB)

frequency ⟶

Figure 11.3 The original blue sine bass has been manipulated with Waves MaxxBass, resulting in harmonic overtones according to the series in Figure 11.2.

Undertones

With some productions you just get lucky, and the low end automatically falls into place. Maybe the key helps, or the type of sounds or the arrangement, but somehow everything fits together. However, the opposite is just as likely to occur: busy kick patterns, sounds with a lot of overlapping frequencies, parts that don't always have their center of gravity in the octave you were hoping for, and so on. This doesn't necessarily mean the arrangement or instruments were badly chosen. A mix like this can also pose a technical challenge, and the result can sound very special if you manage to give everything its own place. In doing so, you may need to rely on more drastic measures than EQ alone.

Sometimes you want to make an instrument play a completely different role in the low end. In these cases, you can use a subharmonic synthesizer to generate energy below the fundamental. This way, you can take a bass part that consists of relatively high notes and make it feel a bit heavier. Or you can position the center of gravity of a busy kick fairly high up the spectrum to make it sound energetic and tight, but at the same time add some extra depth below. It's like combining the best of a high- and a low-pitched kick. Thanks to this approach, instruments that normally don't function as a bass can still take care of the low end. But, just like with other artificial tricks, a little goes a long way. Too much will make the sound sluggish, plus it will use up the energy you badly need for the rest of your mix.

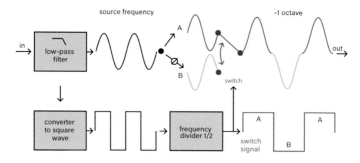

Figure 11.4 A subharmonic synthesizer can generate a signal that's exactly an octave below the input signal's fundamental. The low-pass filter at the input is necessary to distill the fundamental from the input signal, and then turn it into a switch signal. If the fundamental makes large interval jumps, it won't always pass through the filter so well, and the synthesizer can't track it anymore. But with fixed-pitch source sounds like bass drums, the synthesizer's response is usually very reliable.

Filters

So far I've mainly described how you can enhance the sound of the low end, but it just as often happens that there's too much going on in the low range. When this is the case, some elements will have to make way for others. A common misconception in this process is that the lows in your mix should only come from the bass instruments. Some tutorials even recommend you to use high-pass filters on all the other instruments by default, to reserve the low end for the bass instruments. Of course it's true that a double bass has more low-range energy than a trumpet, but this doesn't mean that the bit of low end in the trumpet is unnecessary. So 'if it ain't broke, don't fix it' would be a better approach to using high-pass filters. And if something does get in the way, a minor correction with a shelf or bell filter is often enough to fix the problem, instead of cutting out entire frequency ranges. This focused approach is a bit more of a puzzle and it requires some listening skills, but it prevents the sound of your mix from gradually becoming smaller and flatter.

Are high-pass filters completely unnecessary then? No, absolutely not. Sometimes there really is a lot of unnecessary low-frequency rumble cluttering up your mix. Live mixers know all about this problem. And besides as a tool for cleaning up this mess, some filters can also be used to 'focus' the low end. This requires a filter that produces an amplification (caused by resonance in the filter) right above its cutoff frequency. This resonance bump—which a lot of filters have—was originally a technical imperfection, but it has proven to be surprisingly useful in practice. These filters seem to make instruments larger or smaller. Almost physically so, because the resonance they produce resembles a soundbox or speaker

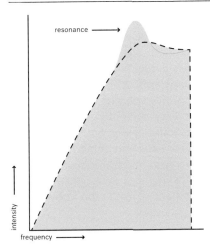

Figure 11.5 A filter with resonance can help to focus the low end: below the cutoff frequency it removes energy, but around the cutoff frequency it adds energy.

cabinet. Instruments that take up too much space can get more low-range focus and definition this way, as you choose one important frequency range and attenuate everything below this range. An example would be when a bass drum has the aforementioned pitch bend, but lacks a clear center of gravity. By making the filter resonate at the frequency where you want the center of gravity to be, the bass drum will excite this resonance with every hit. Again, this is not an effect you will need on all of the instruments in your mix, but sometimes it can add a lot of definition and impact, while still sounding pretty natural.

11.2 The Ceiling: Creating Clear Mixes

High frequencies make sounds feel clear, close, intimate and detailed, as long as they sound good and work together. Making a single sound clearer is not so difficult: you just amplify some high frequencies with an equalizer, and voilà. But creating a clear mix is another story. If you just add highs to your entire mix—or to the sounds it's made up of—you will make it clearer, but not necessarily more defined. This is the result of masking: certain high frequencies in one sound push parts of another sound away. By nature, everyone is familiar with the effects of low-frequency masking: resonating tones, a swampy sound, entire instruments that become inaudible. In the high end the effect is more subtle, but eventually just as annoying. You won't easily lose entire instruments, but your mix will be tiring to listen to, and in spite of its clear sound, it will still seem to be missing focus and perspective.

So our enemy is high-frequency masking. But there's an important warning that comes with this statement: it can be difficult to hear which frequency area causes the perceived masking. Sometimes the problem appears to be in the high end, but there's actually something going on around 300 Hz that makes the mix sound dull. A quick scan with an equalizer can provide a decisive answer. A nice test when you're halfway through the mixing process is to try the 'easy route' by putting a single equalizer on the entire mix and make everything clearer. Sometimes you're lucky, but it's more likely that certain problems will come to light. For example, a voice suddenly seems to have a lisp, or a hi-hat part turns out to be too loud. By solving these problems at the source, you can make the entire mix brighter while keeping it balanced. And who knows: after

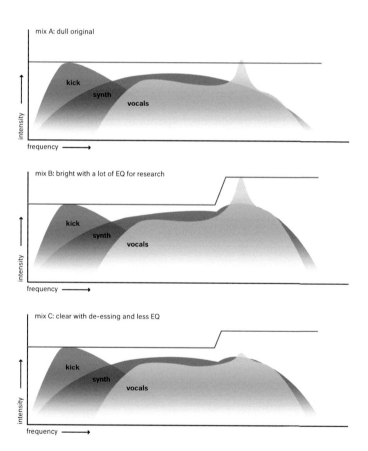

Figure 11.6 Boosting the high end of your entire mix can help to reveal problems that wouldn't stand out otherwise. Thanks to the research conducted in mix B, mix C is much better balanced than mix A, even though the original mix didn't seem to have any notable problems at first.

you're done, you might decide that the mix doesn't need extra highs after all. Even then, it often still helps to solve a few individual problems that would otherwise have stayed under the radar: your mix will be easier to listen to, and it will have a more open sound (see Figure 11.6).

An Upside-Down World

Nowadays it's quite common for sounds to already be very clear-sounding at the source. Digital systems can perfectly capture minute transients, and a lot of recording is done in small (home) studios, where the microphones are often placed close to the source, picking up a lot of detail. And sample library manufacturers usually do everything they can to make their samples sound crystal-clear and impressive right out of the box. The same goes for a lot of synthesizer presets. Surrounded by all that super-defined source material, you'll need to make a choice as a mixer. Some sounds will have to become less defined and more muffled if you want to save the mix. This process can feel pretty illogical, as brighter almost always seems better, and removing definition feels counterproductive.

The best solution is to equalize in the context of the mix. This way, you'll hear the effect on the mix as a whole, and not just the disappearing definition of the sound you're working on. You can imagine how different the thought process would have been in a production of forty years ago. Due to the recording method and equipment, the source material was woolly-sounding, rather than over-detailed. You had to choose which sounds you wanted to emphasize in the high end. This process feels very different than choosing which sound you want to make duller, and it's one of the reasons why some engineers dislike the 'cold sound' of digital.

In the worlds of classical music and jazz, few people complain about the sound of digital. On the contrary: the verdict on modern audio equipment is generally very positive: it's quiet, the resolution is high and it has little distortion. These are all features that come in handy if you want to make a realistic recording of an existing sound image. All of the (spatial) characteristics the final mix needs to have are already contained in the source. Unless you close-mic all the instruments, the mix is actually already made in the recording space. This couldn't be more different in pop music, where you try to compile a brand new sound image out of a variety of separate sources. You'll have to create all the depth and perspective yourself. A good starting point for this is drawing up a priority list for all the sounds in your mix. Is a sound important? Then it can be clear. But if a sound only plays a minor role, it can be duller. You create this role division by trimming the high frequencies. All you need is some experience to know where, when and to what extent you should apply these high cuts. Sometimes this technique doesn't work at all: it always depends on the music and the source material.

Interestingly, a couple of dull elements don't necessarily make a dull mix. In fact, one or two distinctly clear and defined elements can make the entire mix sound clear. An extreme example would be to completely cut the highs above 10 kHz of all the elements in your mix (except the cymbals and vocals). The result can still sound surprisingly 'hi-fi.' If you listen closely to energetic pop music (a genre that forces you to make pronounced mixing choices due to the high density in the sound image), you'll find that the voice often has more energy in the high end than the electric guitars or synths. In the more alternative genres, it's often the other way around: sometimes the lead vocals are downright dull, to the benefit of the rest of the instruments. It's not so much about which element is foregrounded, as long as there's some kind of a hierarchy.

Premeditated Recording

The easiest way to control the high frequencies is by already considering during recording where a sound will be placed in the mix. For example, when you're recording background vocals, you can try to place the microphone a bit further away from the singer than when you recorded the lead vocals. This will automatically make the highs sound milder—and the vocals seem further away. Recording it right will sound much more convincing than placing the microphone close and then cutting the highs in the mix. With the latter approach, you don't just attenuate the highs in the direct sound, but also in the acoustics. As a result, the direct sound will still seem closer than if you place the microphone at a distance. On top of this, you naturally pick up more of the acoustics when the microphone is further away, which makes the direct sound seem even less close. With a lot of effort, this effect can be simulated with short ambiance reverbs, but recording the real thing always sounds more natural.

The problem is that not every recording space is suitable for this purpose. Especially in small studios, things will quickly start to sound 'boxy' if you move the microphone further away from the source. If this is the case, you can also choose to use a microphone that doesn't emphasize the highs, and then place it close to the source. A dynamic microphone might very well be the best choice for close-miking sounds that you want to place in the background of the mix. It's no coincidence that ribbon microphones are currently making a huge comeback. Due to their lack of high-frequency resonances, they sound more natural than condenser microphones with the highs filtered off, if you're looking for a mild sound. This way, your microphone choices can already shape the sound image to a large extent. For example, if you record everything with dynamic microphones and only the vocals with a condenser microphone, this will almost automatically create a lot of room for the vocals in the mix.

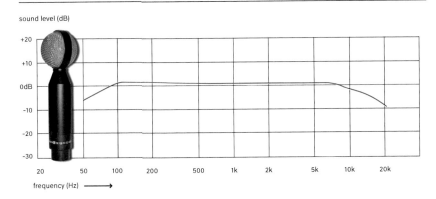

sound level (dB)

Figure 11.7 A ribbon microphone like this Beyerdynamic M 130 can already make high frequencies a bit milder during recording.

Length

The ratio between short peaks and long sounds largely determines whether the highs in your mix will sound open. This is what makes some synthesizers or heavily distorted electric guitars so hard to balance, as these instruments can produce an almost continuous stream of high frequencies. Shakers and tambourines that play busy rhythms aren't much better. It's not easy to make the details of other instruments cut through and fill the gaps without everything becoming shrill and hectic. The longer the decay, the denser the rhythm and the flatter the dynamics of a sound, the softer you have to place it in the mix it to avoid masking other sounds. But placing heavy guitars softly in the mix is usually not what the music calls for.

A bus compressor can help by only allowing the guitars to sound loud in the gaps between other sounds (see Chapter 12), but this method also has its limits. Once this limit has been reached, a low-pass filter on the guitars can provide the solution. There are all kinds of low-pass and shelving filters that can work wonders. But just like in the low end, filters with a boost just below the cutoff frequency are particularly useful for this. The boost gives a sound a slight emphasis in that frequency range, and makes it less apparent that everything above the cutoff frequency is no longer there.

Very steep filters can work well to cut high frequencies, but the results tend to sound unnatural. A less steep filter (6 or 12 dB per octave) usually cuts too much off the midrange before the highs are sufficiently attenuated. However, if you compensate this with a bell boost in the midrange, the result can sound better than a steep filter with resonance. In addition, this method allows you to independently control the attenuation of the highs and the amplification in the range below (see Figure 11.8).

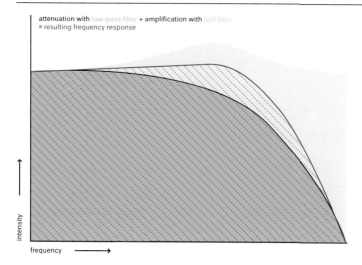

Figure 11.8 A low-pass filter with a mild slope (the orange curve) sounds most natural, but it usually removes more of the midrange than you want. You can compensate this with a bell boost (the blue curve). This approach gives you a lot of control over the resulting filter response (the red curve).

Damage Control

You can only make a clearly defined mix if there aren't any major problems in the music. With a good arrangement, the parts and instruments will already work together without any mixing. And not unimportantly of course: they should be well played or programmed. Especially timing is crucial: peaks that are supposed to coincide, only add up when they're perfectly aligned. Otherwise, you won't hear a joint rhythm, but a weak addition of transients and a lot of clutter. When this happens, you can mix all you want, but you won't make it come together. In that case, there's no other solution than to start cutting, shifting and pasting. But if this is not an option—due to time, money or other restraints—you can pick one rhythmic part as the lead. You can make this instrument clear, loud and powerful, while you attenuate the peaks and muffle the sound of the other parts. This way, at least you won't notice the rhythmically shaky performance so much.

Making Sound . . .

Some pieces of equipment—apart from their function—have a particular sound that affects the high end. This nearly always concerns devices that have a limited frequency response (the highs are rolled off above a certain frequency) or that start to distort above a certain level. In itself, a limited frequency response is not so interesting. It might sound a bit different than a simple low-pass filter, but usually an equalizer also works well for this purpose, plus it's more flexible. So distortion is more interesting in this regard. A tape recorder that saturates high frequencies (if you record them at a high level) can be very useful as an alternative to shrill-sounding peak limiting or conspicuous de-essing. Other useful tools are devices that only distort a specific frequency range, like exciters. The Aphex Aural Exciter is the archetype of this category of processors that add overtones to the high frequencies that you send through. The result sounds very different from using an equalizer: you actually generate frequencies that weren't originally part of the source material. Plus there are some vintage devices with transformers that can't cope with extremely fast peaks, which can help to make sounds less 'harsh-sounding.'

. . . or Breaking It

Besides using gear that adds a desirable discoloration to the sound, it's even more important to prevent undesirable discoloration. Mediocre AD/DA converters with timing fluctuations (jitter), harmonic—but mainly non-harmonic—distortion, clipping, condenser microphones that were intentionally made too shrill (because in a quick comparison, brighter usually sounds better), sample rate conversion and data compression can all throw a monkey wrench in the works. Therefore, it's a good idea to work with a 24-bit resolution and a high sample rate, and to keep an eye on the peak meters on your plugins and in your digital mixer. A lot of plugins use upsampling to beef up their internal resolution, but if this isn't done at a very high quality, it can also clutter up the high end. Some phase-linear equalizers have this tendency as well. The fewer of these plugins you need, the more open the highs in your mix will sound. Thorough comparison is very important: what does a particular processor do to the sound on top of its actual function? You'll soon find out that not all equalizers are made equal. Before you know it, you're very selective about which ones you do and don't want to use to boost the highs. More about the technical background of this can be found in Chapter 17.

11.3 Securing the House: Dynamic Interventions

In section 5.4 you could read about what happens when an instrument suddenly gets too loud. Loud is often a great thing, but too loud means that the illusion of your mix is broken: the instrument that bursts out makes all the other instruments seem small. This problem can easily be avoided with compression, but there's a danger that you need so much of it to solve the problem that the instrument loses its impact and liveliness. Therefore, it helps to focus the compression as much as possible, so as to only affect the sound when it actually needs it. A standard compressor can recognize only one quality: intensity. Below the threshold, nothing happens, but as soon as the sound exceeds the threshold, it's compressed.

This mechanism can be refined with a sidechain filter that makes the compressor respond more (or less) to specific frequency ranges. This already helps a lot to focus the compression on a particular problem. For example, the loud peaks in a vocal track that you want to temper contain a lot of midrange frequencies compared to the softer passages. This characteristic can be exaggerated with the sidechain filter, making it even easier for the compressor to only respond to the peaks. In a way, you're using EQ in the sidechain to tell the compressor what it should pay attention to.

But what if the huge concentration of midrange frequencies in the peaks is in fact the problem? You can make the compressor turn the peaks down as a whole, but this won't get rid of their nasal sound quality. In this case, you actually only want to attenuate the midrange frequencies during the peaks, not the entire signal. There are mix engineers who solve this problem by sending the vocals to two channels, the first with standard EQ settings and the second with considerably reduced midrange frequencies (and a bit more compression). By automating the level ratio between these two channels during the peaks, they can keep the sound of the voice balanced. Sometimes, however, the peaks are so unpredictable —or of such a short duration—that the very thought of automating them is enough to throw in the towel. Fortunately, there's another way to tackle this problem.

Multiband

There are two types of processors that can provide an automatic solution to these 'dynamic frequency issues': multiband compressors and dynamic equalizers. Both can be seen as compressors that only manipulate a part of the frequency spectrum. The difference between the two is that a multiband compressor uses crossover filters to divide the frequency spectrum into a series of adjacent bands, while a dynamic equalizer works like a parametric equalizer that gives you full control over the type of filter and the bandwidth.

Usually, a multiband compressor can only attenuate frequency bands, while a dynamic equalizer can also amplify them (see Figure 11.9). Which of these two processors is most suitable for you depends on the application. If it's your goal to compress the entire signal, but for instance with a lower ratio and faster timing for the higher frequencies, then multiband compression is the best solution. But if you want to focus on a specific problem, such as loud S sounds or acoustic resonances, a dynamic equalizer is probably your best option. Some dynamic equalizers allow you to perform pretty advanced tricks, because—unlike a multiband compressor—the filter's sidechain doesn't only have to contain the corresponding frequency band of that filter. For example, you can feed the dynamic filter's sidechain with the low frequencies of the signal while the filter itself is manipulating the highs. This way, you can boost or dip the highs in a drum mix every time the bass drum hits. Some processors also allow you to use an external sidechain signal. For example, you can attenuate the midrange of the guitar only when the vocals are present in the sidechain.

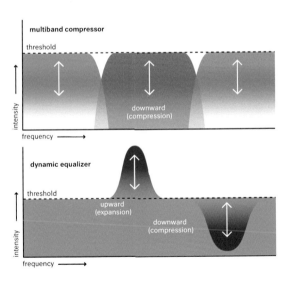

Figure 11.9 A multiband compressor always uses crossover filters to divide the audio into bands, regardless of whether compression occurs. Most dynamic equalizers use parametric filters, which only affect the audio during compression or expansion.

De-essers

De-essers are dynamic processors specifically designed to reduce excess S sounds in vocal recordings, but they also come in handy if you want to tackle high frequency peaks in various other sounds. There are two types of de-essers. The simplest version works like a compressor with a high-pass or band-pass filter in the sidechain. This de-esser attenuates the overall signal when an S sound triggers the processor. This type generally sounds very pleasant (when used lightly, of course). If more drastic measures are needed, the second type often works better. This is a dynamic equalizer with a single frequency band, which you can set to the problem frequency. Because only the problem frequency is attenuated, you can remove the excess sibilance from a sound, with the risk that the result will sound unnatural.

The nicest variant of this design works independently of the incoming signal level. This means there's no fixed threshold from where the de-esser starts to work, but a kind of scale that measures the ratio between the high and low frequencies. As a result, the device only intervenes at moments when there are more high than low frequencies. This way, you can also tackle soft lisping, without the de-esser overreacting if the singer suddenly bursts out afterwards.

Disclaimer

Multiband compression is so flexible that you might think you want to use it all the time, but there are two reasons why that's not a good idea. First of all, the use of crossover filters also has its drawbacks: it's detrimental to the integrity of the signal. But the second reason is even more important: the well-known adage 'if it ain't broke, don't fix it.' The essence of mixing is the creation of musical relationships: the way different instruments relate to each other. Each of these instruments is part of the arrangement for a reason, and also played or programmed in a certain way for a reason. A cello playing staccato accents sounds much sharper than the mild legato passage that preceded it, and so it should. Even if you decide to compress the cello because its dynamic range is too wide to place it in your mix, the mild and sharp parts will still sound very different. Therefore, in spite of the reduced dynamics in terms of signal level, you will still perceive a lot of musical dynamics. (This is also why it's important to capture an almost exaggerated amount of expression in a recording, as this makes it easier for this recording to stand out in the big picture; it can withstand some compression without losing the sense of dynamics.)

However, if you use a multiband compressor to reduce the dynamics, the entire sound of the cello will be evened out. Harsh peaks become less harsh, and mild passages less mild. So it's important to be careful with multiband processing: before you know it, you're removing all the contrast and expression from your mix. Multiband compression can be a great problem-solver for individual sounds—better than any standard compressor or equalizer out there—but if you get to the point where you start applying it to your entire mix, it's time to stop and think about what you're doing. Usually, it's a better idea to individually analyze the problems that brought you to that point, and solve them in their own respective channels. If you don't do this, you can easily end up using multiband compression to even out all the minute contrasts that actually made your mix exciting. It takes a bit more research, but this focused approach will ultimately result in a mix with more impact.

Dynamics without a Threshold

The attack and release setting on a compressor can be used to attenuate or emphasize the attack or sustain of a sound. It's a great way to maneuver sounds to the fore- or background of your mix, but if you want to completely change a sound's character, you usually need so much compression that the sound is audibly 'restrained' and ends up sounding small. Therefore, the next step is a processor that doesn't see dynamics as the difference between loud and soft, but as the difference in the speed with which the signal gets louder or softer. SPL were the first manufacturer to market such a device, and the name 'transient designer' has become synonymous with countless plugins and hardware based on this idea.

It works like this: the signal is split into an attack part and a sustain part. This division is made based not on a fixed threshold—like a compressor would do—but on the change in signal level. This means that the process works independently of the absolute signal level (see Figure 11.10). After the split, you can amplify or attenuate the levels of the attack and sustain parts separately.

But how can this be used in practice? In many ways. For example, you can only make the attack of a snare drum sharper, regardless of how loud it's played. Or remove the sound of the room from a drum recording, without having to use a gate that only opens when the drummer plays loud enough. Or you can do the opposite, and boost the acoustics in a recording. You could also use heavy compression for this, but a transient designer can do it without squashing all the peaks. The effect works

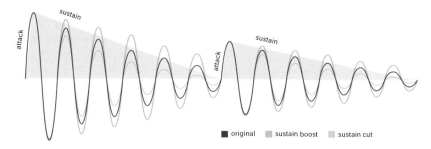

Figure 11.10 Using a transient designer to manipulate the sustain of a signal. The signal level doesn't affect the process, as the processor only looks at the slope of the signal's envelope.

particularly well with percussive material, but other sounds with a clear attack–sustain division (like pianos or acoustic guitars) also respond well to this kind of manipulation. I've even managed to make hollow-sounding recordings of string quartets dryer with a transient designer. Some transient designers can perform their manipulations frequency-dependently. For example, they can only remove the low-frequency decay from a sound and leave the highs alone. It's like you're adjusting the acoustics of the recording room, but after the sound has already been recorded.

Closing the Gates

A significant part of the mixing process consists of removing unnecessary information. Besides certain frequency ranges that can get in the way, there are also a variety of noises to annoy you. For example, an instrument microphone that has picked up a lot of spill from other instruments, but also breathing and saliva sounds in a vocal recording, or electronic and ambient noise. Back in the day when most music was still recorded on tape, it wasn't as easy as it is now to simply turn off those unnecessary sounds. The visual feedback and precise non-destructive editing that has become the norm didn't exist yet. That's why expanders and gates were very popular tools.

Expanders come in two flavors: upward and downward. An upward expander can be seen as a reverse compressor: signals that exceed the threshold aren't attenuated, but amplified according to an adjustable ratio. Conversely, a downward expander attenuates the noise that remains below the threshold according to an adjustable ratio (see Figure 11.11). A gate is a downward expander with a very high ratio, so it only allows sounds to pass through that exceed its adjustable threshold level. This makes it ideally suited to automatically mute the noise of a guitar amplifier at the moments when the guitarist doesn't play. Just like compressors, gates and expanders have attack and release parameters to adjust their reaction speed. When you use a gate on drums, it should be able to open extremely fast, since you don't want to attenuate the first transient. This is why some digital gates have a 'look-ahead' function, a short buffer that allows them to look into the future and always open in time. To make the triggering of the gate even more reliable, many gates have sidechain filters to make them respond to a particular frequency range.

A gate is a pretty uninteresting processor, except if you use it for less obvious purposes. Some gates open so suddenly that they can really add some bite to the attack of a sound. Or the opposite: if the transients of a sound are too sharp, a gate with a long attack time can round them off a bit. Especially if you want to shape the envelope of continuous sounds it's necessary for a gate to have an adjustable attenuation range, as the sound would be completely switched off otherwise. An attenuation of 6 dB when the gate is closed is often effective enough, without sounding unnatural. If the effect still isn't smooth enough, it's better to use an expander, so you can lower the ratio and maybe use a soft-knee setting. But unlike those of a transient designer, the manipulations you make with an expander are still level-dependent.

Perhaps the most fun applications for gates are those in which the opening of the gate isn't controlled by the input signal, but by a completely different signal. You can use a gate with an external sidechain input

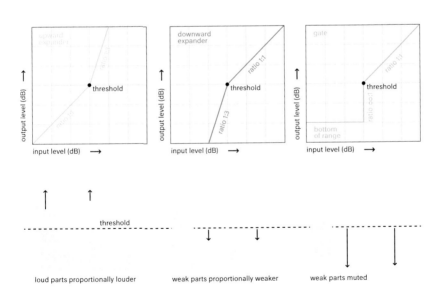

Figure 11.11 Typical response of expanders and gates. Some gates are also downward expanders, and vice versa: if the ratio is adjustable, you can create both response curves.

(which is sometimes called 'key') to only turn on the room mic of a drum recording when the snare drum is hit. And if you don't care so much about the sound being natural, you can also create all kinds of musical connections this way. The best-known example is a drum loop that triggers a rhythmic stutter in a static pad sound (see section 9.5 for more on sidechain triggering).

Hard Work

In spite of everything that automated processing has to offer, in many situations there's still a strong case for old-fashioned handicraft. Using a de-esser to intervene at five annoying moments in a song can be detrimental to the overall sound of the vocals. If you look at it this way, it's a small effort to turn those five lisped S sounds down by hand (or locally attenuate the highs) and leave the rest of the recording alone. The same thing goes for gates: if a drummer only hits the toms during a couple of fills, it's much quicker and more reliable to single out these few hits, give each of them a natural-sounding fade in and out, and mute the rest of the audio.

11.4 The Exterior Walls: Stereo Processing

At first sight, mixing in stereo is much easier than in mono. You have much more room available to give each part its own place, so the sounds don't need to overlap so much. It's easy to think you're doing a good job when you're sitting in your 'mixing lab,' in a perfectly equilateral triangle with two good speakers. But the result can be very disappointing when you hear it somewhere else. Sometimes your entire mix falls apart, while on other stereos it sounds way too woolly and undefined. This is because you rarely come across a good stereo arrangement 'in the wild' (see Figure 11.12). But if you design your stereo image in such a way that it can

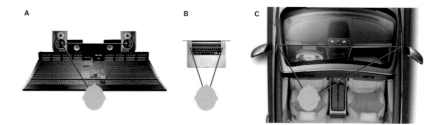

Figure 11.12 When it comes to stereo reproduction, there's no place like the studio (A). On a laptop or docking station, the sound is almost mono (B), while in a car it's too wide and lopsided (C).

withstand extreme conditions, you'll have a higher chance of success than by starting a crusade against bad speaker placement.

The most robust mix you can make is in mono, as this means that all the elements are equally reproduced by both speakers. Such a mix can never be unbalanced if one speaker is closer or louder than the other. It won't fall apart if the speakers are placed too far away from each other, and it won't clog up if they're too close together. The only thing is that it will sound a bit one-dimensional. Still, it's better to make sure that the important elements in your mix are (about equally) present in both speakers. This way, at least these parts will always be heard clearly, regardless of the listening situation. Then you can use the remaining elements, various reverbs and any room microphones or effects to make the overall sound more spacious. The result is that in locations with subpar speaker placement you might lose the icing on the cake, but the essence of your mix will still be intact. And if you know what you're doing, it can still sound very impressive and spacious through well-placed speakers or on headphones.

Sum and Difference

What it comes down to is that there shouldn't be too many differences between the left and right channels, but the few differences there are should have a strong effect. This isn't always easy to achieve, so it can help to manipulate the differences and the similarities separately. A mid/side (M/S) matrix can split a stereo sound, stereo group or your entire mix into sum and difference components. After this split, you can change the levels of these individual components, or manipulate them with effects like EQ or compression (see Figure 11.13).

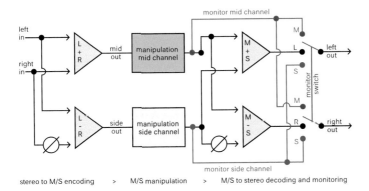

Figure 11.13 An M/S matrix can be used to convert stereo to M/S. This allows you to manipulate and monitor the M and S channels separately (with the monitor switch) and then decode the result back to stereo.

Nowadays, many plugins already have an M/S matrix built in by default. But there are also separate plugin matrices that allow you to perform any kind of manipulation (digital or even analog through a DA/AD loop) in the M/S domain. You place one of these matrices in encoder mode before the plugin or hardware that you want to use in M/S, and then another matrix in decoder mode to turn the manipulated sound into stereo again. Of course it's important that your equipment can work in dual mono and not only in stereo link, since you want to manipulate the M channel independently of the S channel. One warning though: manipulations that cause phase shifts or delays can seriously disrupt the relationship between the mid and side channels, which means that the decoder won't be able to reconstruct a decent stereo image. This is why you should use EQ sparingly, and make sure that you always compensate for any latency.

The name mid/side is a bit unfortunate, since it's not exactly a division into what's in the middle and what's on the outer edges of the mix. In reality, the M component is a sum of left and right. This means that it contains all the shared signals, but also the signals that are only present in one of the two channels (like hard-panned mono sounds). The S component is created by subtracting the right channel from the left. This is done by adding the right channel to the left, but in reverse polarity. The result contains all the differences between left and right, and again also the signals that are only present in one of the two channels. So the system isn't a perfect tool for separating mono and stereo information, but still very useful. Plus it can give you interesting insights into the structure of

---

Phantom Center

There are also processors that—unlike M/S matrices—isolate only the sound that's located straight in the middle of the mix (the phantom center). With these systems, hard-panned elements (which are only present in one of the two channels) don't end up in the mid component. As a result, these processors are better suited for mastering manipulations like turning the vocals up or down while leaving the rest of the mix alone. This sounds like a dream come true for wedding planners who want to make karaoke versions of old singalongs, but these processors also have their drawbacks. They affect the structure of the audio they manipulate much more deeply than M/S matrices, and these structural changes can cause sonic side effects. But sometimes they do offer a solution in cases where M/S is found lacking.

your own mixes and your reference material. Once you've separated the M/S components, you can also listen to them individually. In Figure 11.13 you can see there's a monitor switch that lets you choose what is sent to the outputs. Many M/S matrices have such a feature, which you can also use to analyze your music collection. By only listening to the M or S component, you can learn a lot about the 'DNA' of a mix. You'll especially gain insight into the sonic material that produces the sense of size and stereo width—and how this material sounds exactly. If you then apply this knowledge to your next mix, chances are you won't need any M/S manipulation in the end.

Manipulation

Why would you go to the trouble of manipulating a stereo mix in M/S while mixing, if you can influence the stereo image much more accurately by panning individual components or manipulating their levels? Not necessarily because one method leads to better results than the other. The biggest difference is that in M/S mode you can simultaneously control various underlying parameters. It feels like you're manipulating the overall spatial characteristics of the mix, rather than small aspects of it. You can immediately hear the effect of your manipulations on the sound of the entire mix; it works intuitively. But you can't make big changes, as the sound image would quickly fall apart. A nice compromise is to first draw the big picture of the mix, and then use M/S processing for fine-tuning.

When you reach the final phase of your mix, most of the time you'll already have arranged all the elements in stereo groups. For instance, a group with stereo drums and bass, a group with stereo chord instruments and a group with mono vocals. It often happens that at the last moment there's still some friction between frequency ranges, so the mix doesn't really fit like a puzzle yet. For example, the chord instruments still get in the way of the vocals too much around 2 kHz. But if you dip these instruments at 2 kHz, they will sound weak and lifeless.

Since the vocals are in the center of the mix, they naturally clash the most with the chord instruments that also occupy that area. Therefore, you might be able to fix the problem by only attenuating 2 kHz in the M component of the chord instruments. This way, you will create more space for the vocals, but without changing the overall sound of the chord group all that much (see Figure 11.15).

Another possible solution is to simply turn down the chord group, and at the same time use an M/S matrix to boost the S component a bit. This will make the entire group wider, so it will 'wrap' around the vocals more easily. Just make sure your mix doesn't rely too heavily on these constructions, as it won't translate so well to places where the stereo setup is less than ideal.

Vinyl

Initially, mid/side processing didn't have a creative, but a purely technical function. In vinyl cutting, it was (and still is) important that the amount of stereo information in the signal isn't too large. In other words: the left and right channels shouldn't be too far out of phase. Otherwise, the two sides of the groove can move so close toward each other that they push the stylus (needle) out of the groove. In order to overcome this problem, vinyl cutting engineers used a limiter to separately restrict the M and S components of the signal. This was called vertical/lateral processing, due to the direction of the groove. This technology could also be used to filter the lows out of the S channel, making sure that large low-frequency amplitudes were confined to the M channel.

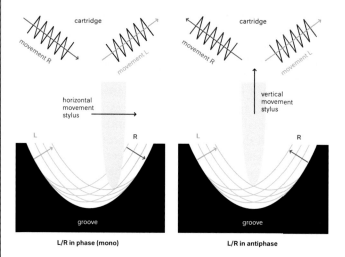

Figure 11.14 By separately registering the two movement directions of the stylus in the groove, the cartridge can distinguish between the left and right channels. This works well, except when low frequencies (with large amplitudes) are in antiphase in both channels: the groove can then become so narrow that the stylus jumps out.

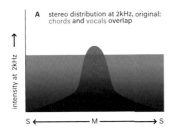
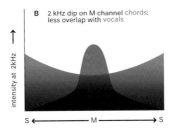

Figure 11.15 M/S filtering can be used to only EQ the mid component of the chord group, creating space for the vocals.

Widening

A wide stereo image is impressive, so it's hardly surprising that there are countless processors (unfortunately even in consumer electronics) that can stretch the stereo image to extreme widths with one push of the button. Most of these processors are based on M/S processing. Many simply boost the S channel, but some amplify the highs in the S channel to highlight spatial details and/or the lows to make the mix seem much bigger (this is also called shuffling). All of these manipulations can easily be done in the mix as well, with an M/S matrix and an equalizer. But no matter what devices or plugins you use, it's a good idea to keep an eye on the correlation between the left and right channels. You can do this by checking the mix in mono every now and then, and by using a correlation meter (see Figure 11.16). This meter indicates the extent to which the left and right channels correspond (see section 7.1 for more on phase correlation). If it regularly displays negative values, you know you probably went too far, making the translation of the mix unreliable.

M/S isn't the only technique that you can use to stretch the stereo image or turn a mono sound into stereo. Just think of all those modulation effects (chorus, vibrato, flanging, phasing, tremolo) that apply slightly different manipulations to the left and right channels. Or constructions with panned short delays. Basically, any method to make a copy of a signal that sounds different from the original can be used to generate stereo information. One of the easiest techniques—which also sounds a lot more natural than a 'washing machine' chorus or a 'fighter jet' flanger—is filtering the left and right channels differently. For example, you can boost the highs a bit more on the dominant side of a sound to subtly emphasize the placement. But you can also create complex cross-over patches that divide the sound into a number of frequency bands, and spread these across the stereo image. If the slopes of the filters you use aren't too steep, the phase distortion will remain within bounds, and the result will even be perfectly mono compatible.

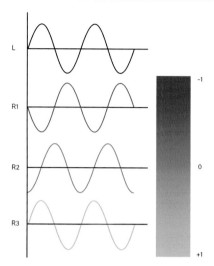

Figure 11.16 A correlation meter indicates the degree of similarity between the left
and right channels of your mix. If you often see it hovering around
zero, or even going into the negative part, there's work to be done.
This figure shows three versions of a right channel measured against
a fixed left channel, and the resulting correlation values.

Have it Both Ways

In EDM and many other electronic genres, the impact of the low end is
paramount, so you don't want to waste energy by allowing the two
speakers to reproduce different things. Only if they work together by
being perfectly in phase, they can deliver the maximum amount of low-
range energy to the listener. But what if the bass synth also takes the role
of the lead, which should preferably sound big and wide? In that case,
you can use a crossover filter to split the sound in a low and a high part.
The lows can then stay mono (or turned into mono if the sound was
already stereo), while the midrange and high frequencies can be
manipulated as much as you want in terms of panning, widening, reverb,
and so forth.

It's remarkable how much of a determining factor the higher range of
the frequency spectrum is for stereo placement. If this range sounds wide,
it will be much less apparent that the low end is mono. Only on headphones you'll notice that it doesn't sound as massive as you had hoped,
because you don't perceive any low-range phase difference between
left and right. You can solve this problem by keeping the core sounds
mono in the low end, and then adding a short ambience reverb, or room
microphones with amplified lows. This way, you'll have a tight and
defined foundation that will still sound wide enough on headphones.

11.5 Beyond the Exterior Walls: 3D Panning

Stereo can sound great, but it's still a kind of peephole diorama: you hear a scene taking place through a limited window. This isn't very different from going to a pop concert where everything also takes place on a stage in front of you (and sometimes even with mono sound). There's one major difference though: at a concert, the acoustics of a large venue reach your ears three-dimensionally, versus the two dimensions of listening to a stereo recording of the same concert. Plus you hear the acoustics of the living room around you, which gives a whole new context to the sound. There's nothing wrong with that—it even makes listening on speakers a lot more natural and relaxing than on headphones—but it does conflict with the image you're trying to conjure up. It would be nice if you could project the recorded acoustics around you like a hologram, while the instruments stay between the speakers. Of course you can do this by recording and mixing in surround, but this will significantly limit your audience. Since the majority of music fans listen in stereo (through speakers or on headphones), it pays to push this system to the extreme.

Hearing in Three Dimensions

It's actually pretty strange that we only have two ears and not three. A third ear would make it very easy to locate exactly where a sound comes from. With two measurement points, you can only trace the horizontal angle of a sound, whereas three points would allow you to also distinguish the vertical angle, and make a distinction between front/back and above/below. This is impossible with two measurement points: two mirrored sound sources placed at exactly the same distance from the listener, one in front and one behind, will be perceived as identical (see Figure 11.17). Still, your ears are hard to fool, because their measurements are supplemented with additional information. For example, the acoustics of your auricle create a filter that sounds different for each angle. And our visual perception also helps to support what we hear. But one factor that shouldn't be underestimated is movement: when you turn your head, the time shift and change of timbre between your two ears immediately tell you if a sound is coming from above, below, in front or behind. For the same reason, moving sound sources are easier to locate than stationary ones.

Binaural Sound

Except for head movements, all aspects of three-dimensional sound can be captured in a recording with a dummy head with two pressure microphones

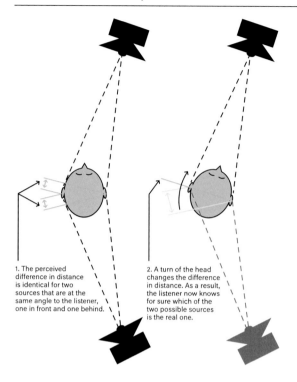

1. The perceived
difference in distance
is identical for two
sources that are at the
same angle to the listener,
one in front and one behind.

2. A turn of the head
changes the difference
in distance. As a result,
the listener now knows
for sure which of the
two possible sources
is the real one.

Figure 11.17 Just by moving their heads, listeners can reliably distinguish between front, back, above and below.

in the ear canals. If you listen to such a binaural recording on headphones, it's like you're experiencing the recording situation again. That is, until you turn your head, because then the sound image will simply turn along. This is why it's difficult to distinguish between front/back and above/below in these recordings. Especially with stationary sound sources, the effect won't be more than a general illusion of spaciousness, without a clear sense of placement (even though it already sounds much more realistic than a normal stereo recording). The superlative of this is binaural synthesis. Combined with head tracking, this system can generate a binaural mix from a surround mix. This is done by superimposing the time and frequency response of both ears for a particular angle of incidence (the head-related transfer function, or HRTF) over the signal coming from that angle. The system then keeps track of the listener's head movements and adjusts the synthesis accordingly. This means you can turn your head all you want, but the sound won't move along with you!

Attenuating Crosstalk . . .

So, combined with the tricks of synthesis and head tracking, binaural sound is the key to recording and playing back three-dimensional sound. As long as you manage to deliver the exact right signal at the right moment to each ear, the illusion will be convincing. However, this is not so simple (if you can still call it that) if you want to use speakers instead of headphones, especially when you only have two at your disposal. This is because the big difference with headphones is crosstalk: the sound of the left speaker also ends up in your right ear, and vice versa. On top of this, the timing and level ratio between left and right are distorted if you're not exactly in the middle between the two speakers, which will blur the sound image even more.

The first problem is pretty easy to solve with a crosstalk canceller, which sends an extra signal to the right speaker that's in antiphase with the left speaker signal, to attenuate the crosstalk of the left speaker to the right ear. And vice versa, of course (see Figure 11.18). This only works for lower frequencies though, because the short wavelengths of high frequencies require too much precision in terms of listener placement. In other words, it would only work if your head is stuck in a vise. Fortunately, your head is a pretty good barrier to high frequencies, so by nature, crosstalk isn't as much of a problem in the high end of the spectrum.

Still, this doesn't solve the second problem: the distorted relationship between left and right when you're not exactly in the middle. In fact, it's even exacerbated by the filtering that's added to the mix. You can clearly notice this with systems like QSound, Roland's RSS or binaural panning plugins. As soon as you move away from the sweet spot, the sound

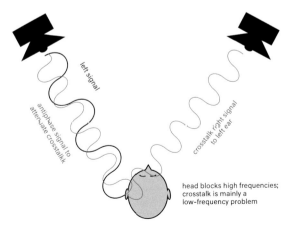

left signal

antiphase signal to attenuate crosstalkk

crosstalk right signal to left ear

head blocks high frequencies; crosstalk is mainly a low-frequency problem

Figure 11.18 The principle of crosstalk cancellation.

becomes unnatural: much worse than a regular stereo mix. The only way to tackle this problem is with a signal that's specially adjusted to the listener's movements. And though there are even systems that use webcams to register what the listener does—and generate the necessary HRTF coding and crosstalk cancellation for this—it's pretty discouraging to think that the acoustics will then heavily mangle these signals anyway. For this reason alone, truly realistic three-dimensional sound reproduction with two speakers will probably remain an illusion.

. . . or Not?

Often, you don't need realism to convey a musical message. Credibility is another story, but by using a pair of speakers to reproduce music, you accept that you're letting go of the reality of a live performance. If you look at the problem of three-dimensional sound this way, it becomes less important to create a perfect hologram, and instead it's more about a general sense of spaciousness. By setting the bar a bit lower than those manufacturers who tried to reproduce 'real 3D sound' through speakers, your mix will sound much better when you're not exactly at the sweet spot. Without crosstalk cancellation, and with much milder filtering than HRTF's extreme (comb) filters, you can create a mix that will sound wide and deep through speakers and on headphones. This way, instead of trying to make a real binaural mix (which only sounds good on

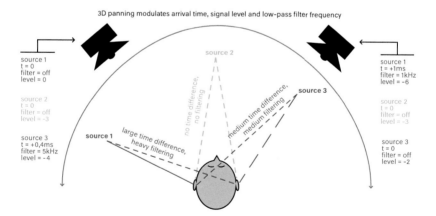

Figure 11.19 Through the use of time differences and low-pass filters, you can give sound sources a convincing place in the stereo image, even beyond the width of the speakers. However, it's still difficult to transfer these time differences to the ears through speakers. Exaggerating the amount of time difference, intensity difference and filtering helps to make the placement a bit less dependent on the listener's position.

headphones), you're actually emulating a recording with two spaced microphones. Between these microphones, there's a small barrier that blocks some of the high frequencies (acoustically, this is called a Jecklin disk). This method, which is the simplest way to simulate a head, can create a pretty convincing illusion without coloring the sound too much (see Figure 11.19). If you're curious how this might sound, take a look at the discography of engineer Tchad Blake.

By experimenting with the amount of time difference and filtering, you can find a balance between effective placement and a wide sweet spot. The more time difference and filtering, the more distinct the placement and the wider the sweet spot becomes. On top of this, big time differences will reduce the comb filter effect when you listen to the mix in mono. That is, until you reach the point where the time difference is so big that left and right are completely disconnected, and the phase difference that's supposed to 'push' the source sound outward from the other side isn't there anymore. Plus the center of the mix falls apart, an effect that's similar to placing your stereo microphones too far apart during recording.

Distinction

With the panning method described in Figure 11.19 you can push sounds all the way to the side, and even place them beyond the width of the speakers. This will create more room in the stereo image for sounds positioned between those extremes. However, this method does cost more energy in the mix than simply panning a mono source hard to one side, as the delayed copy on the other side can also get in the way of other sounds over there. Therefore, it pays to make clear choices. For instance, take one or two elements that can be distinctly wide, and use regular panning for the other sounds. A mix feels as wide as the widest element, so you won't gain that much extra width by moving a whole bunch of sounds to the extremes of the stereo image. Through selective panning, you can create very big contrasts in terms of placement, resulting in a very detailed mix.

Acoustics

If there's one element that's uniquely suited to be made extra wide, it's acoustics (either recorded or subsequently added). Ideally, the listener would be surrounded by the acoustics, just like in a live situation. It's a common misconception that low frequencies are unnecessary in acoustics, because they would only get in the way of the instruments. Above all other things, it's the phase differences between left and right that evoke a sense of size and width, and the sweet spot for perceiving these is only large for low frequencies. In the case of a loud bass guitar, these phase

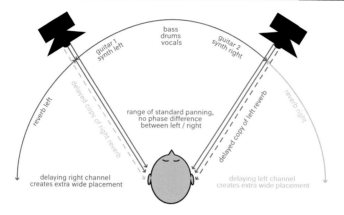

Figure 11.20 A compromise between standard panning for the source sounds and 3D panning for the reverb. This can produce a very convincing illusion if you're in the sweet spot or listening on headphones, and the mix won't fall apart if you hear it on a laptop, for example. Not all reverbs need widening using the illustrated cross-panned delay, some are already designed to sound outside the speakers by default.

differences should probably be avoided: it's a waste of energy and it ruins the mono compatibility of your mix. But reverb takes place at a much lower level in the mix, and it actually complements mono sources (see Figure 11.20). Just give it a try, and turn off the high-pass filter on your reverb. If this makes the sound muddy, don't filter out the deepest frequencies, but rather trim the low midrange of the reverb and shorten the reverberation time.

Once you get used to this method of adding size to your mix, you're bound to discover the added value of short ambience reverbs. You'll immediately understand why certain unrealistic-sounding devices are still very popular. A reverb that sounds unrealistically wide and hollow on solo can actually work great in a mix. It's no wonder that the designer of the classic Lexicon reverbs, David Griesinger, is such a strong advocate of stereo subwoofers: phase differences in the deepest lows are the key to three-dimensional sounding stereo mixes. And if people end up listening to your mix in mono on an iPhone speaker, only the spaciousness will be lost, not the instruments: a pretty good compromise.

Movement

A moving sound source makes everything more convincing. As a listener, you don't need to move your head to determine the direction a moving sound is coming from; this is only necessary for static sources. Movement is an application that HRTF panners excel in. With these panners,

Stereo for Headphones

Whether you use headphones when mixing or not, you can be sure that people are going to listen to your final product this way. I personally like the idea that my work is heard on headphones, because it means that listeners will hear a lot of the details that I heard in the studio. The stereo image will be different of course (it is 'stretched' in a way, and the elements in the center are a bit too loud), but usually this isn't really a problem. In any case, I try to keep the stereo image balanced and not make the panning so extreme that it starts to distract. Often this is enough to make a mix sound good on headphones. Technically, this could be the end of the story: you've created something that works well on speakers, and it's also suitable for listening on headphones. But can't you do something extra to enhance the experience, especially for all those headphone users?

The Devil Is in the Detail

Even if you're not going for a headphone-specific binaural mix for the reasons mentioned above, there are some garden-variety tricks that you can use to create effects that work particularly well on headphones. For example, it's never a good idea to send two halves of an important sound with opposite polarities to the left and right channels of the mix, but this rule doesn't apply to effects and other soft, supplementary sounds. Turning one channel down and reversing its polarity will give such a sound a huge width and move it outside the stereo image, leaving more room for other elements in the mix.

And since we're talking about turning things down: the advantage of headphones is that you can hear so much detail. But this also means that the detail should be there, otherwise your mix will sound too empty. If there's not enough detail in the sound, you can also add some hidden elements to keep the mix interesting. Just think of short echoes or other effects. As long as you turn them down enough, you can often get away with the craziest things that subconsciously add a certain atmosphere to the sound, or that keep the mix moving. This extra 'ear candy' also helps to compensate for the lack of impact on headphones, compared to speakers with the volume turned up.

you're not listening to a static filter effect—the success of which depends heavily on your listening position—but to a modulating filter, so the illusion is much clearer. In a way, the sound change draws your attention to what's going on, while the effect can also remain unnoticed if it's stationary. Plus a moving comb filter is much less recognizable as unwanted coloration than a stationary one—which clearly affects the character of the sound. If you really want to go overboard with this, you can even emulate the Doppler effect. The combination of a pitch bend, a filter sweep and volume automation can suggest a spectacular movement. The only problem is that a pitch bend is usually out of the question with tonal material, unless you manage to turn it into a musically useful glissando of course!

Chapter 12

Bus Compression

The Sum of the Parts

When I first started in my home studio, the bands I recorded were not made up of professional musicians. The consistency in their playing (as well as my own by the way) often left a lot to be desired. I soon learned to use compressors to keep the parts together a bit more. For a long time, a compressor remained just that in my mind: a tool to make sure that no notes are too loud or too soft. Later, when I started working with electronic music, I learned how others mix in that genre. To my surprise, all kinds of compressors were used. For years I had seen compression as a tool to provide a more consistent snare drum sound to inconsistent drummers. But now I suddenly discovered that people even applied compression to perfectly even-sounding drum samples. Why on earth would they do that?

I knew you could use compression to completely distort a sound and give it a different character. But it required new insights to understand why the producers I worked with also used compressors on groups of instruments (buses), and often on the entire mix as well. I started to try it myself and, for the first time, I heard a completely different effect of compression: it can bring together individual parts.

12.1 Influence

Open arrangements might seem easy to mix at first. Take for instance a pop song with just a kick, a short bass note and the vocals playing on the beat, and all the additional elements only on the offbeat. In an arrangement like this, the lows get plenty of room and they won't easily get in the way of other things. But if you think you can easily make this mix sound impressive, you could be in for a disappointment. Low frequencies mainly seem impressive when they affect other elements, like a thundering explosion that drowns out everything else. A perfectly defined mix that lacks this kind of drowning out sounds weak. But turning the lows all the way up is usually not an option in a studio production, as this only works when you play the mix at thunder level. The mix should still be impressive at a low volume, so another solution is needed.

The drowning-out effect can also be created artificially with compression. If you send the bass instruments relatively loudly into a bus compressor that's on the entire mix, the compressor will mostly respond to the lows. This way, the entire mix is pushed back at the moments when there's a lot of low-end energy. And if you don't set the attack too fast, the initial punch of the lows will still be allowed through. The trademark EDM sidechaining of the kick drum takes this to the extreme. The kick triggers the sidechain of an extreme compressor, which is applied to all the other elements in the mix. This way, it ducks the entire mix, while the kick itself stays intact. As a result, it will even seem relentlessly loud on laptop speakers.

A compressor will always be a device that makes loud things softer, whether you put it on individual sounds or on an entire mix. The main difference is in the interaction between sounds that results from running them through a compressor as a whole. The sound that makes the compressor work the hardest also compresses all the other sounds in the mix. Especially in music that's composed of isolated elements, this method can make it sound more like a whole. An orchestra that's recorded live rarely needs such a treatment, because the acoustics and crosstalk already provide enough cohesion between individual parts. But if you're working with orchestra sections or individually recorded samples, bus compression can make the resulting mix sound more like one orchestra playing

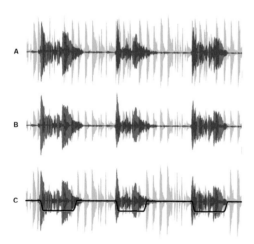

Figure 12.1 The vocals (red wave) and guitar (green wave) don't blend well. They're either too weak (A) or too dominant (B). If you use a bus compressor that mainly affects the vocals (because they're relatively loud at the input), you can make the guitar relatively weaker when the vocalist is singing, and let it come back up at the moments in between—while simultaneously controlling the overall dynamics (C).

Bus Compression Is Not a Cure-All

This chapter is deliberately not an addition to the earlier chapter on dynamics and compression, because bus compression can only add something to a mix that already works. It's about subtleties that will subconsciously make the mix feel better, not about radical changes. The effects discussed in this chapter shouldn't be noticeable to the average listener. Therefore, if you need so much bus compression to make your mix work that the effect starts to draw attention to itself, it's usually a good idea to first refine the basic balance.

together. In a lot of pop music, this type of compression has partly taken over the role of reverb and acoustics: the softer and longer sounds almost seem like reverberations of the rhythmic and loud sounds, due to the fact that the compressor mainly brings them out between the loud notes. So just like reverb, bus compression can serve as the glue that holds sounds together (see Figure 12.1).

Border Patrol

Besides as glue for your mix, a bus compressor can also be very interesting as an indicator. You can see the compressor as a kind of soft ceiling, guarding the borders of your mix. When you're struggling with mix balances and find yourself going down that treacherous road where you're only making everything louder, it can function as a counterbalance. It forces you to listen more carefully where the clashes in the arrangement occur: where does the energy pile up, which parts get in each other's way, which instrument is the first to 'choke' the compressor, and so on. If you get rid of these clashes with EQ and stereo placement, the need to keep pushing things up will decrease. And maybe you won't even need a bus compressor after this, since it has already done its job as an indicator.

This problem indication also takes place on another level: if you notice that your mix responds badly to bus compression, it probably won't work well in listening situations outside the studio. Apart from the obvious radio compression (which is a lot more relentless than your own bus compression), a listening environment with less-than-ideal acoustics can also create a compression-like effect. In a way, acoustics are like a binding agent that keeps the sound together, but if your mix lacks definition, details can easily disappear in this 'resonant soup.' Listening to the effect of bus compression on your mix invites you to create enough

Bus Compression to Increase Dynamics?

The reason for using compression on groups of instruments or an entire mix can be somewhat puzzling at first. It seems much less flexible than controlling the dynamics of individual components directly. Because everything is interrelated, the negative side effects of compression can make an obvious appearance across multiple instruments, which makes setting up bus compression a delicate task. But once you familiarize yourself with this procedure—and

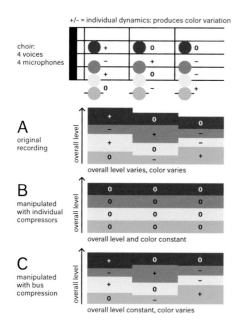

Figure 12.2 Two ways to control the varying dynamics of a four-voice choir part, which was recorded using close microphones on each of the voices. Each voice is depicted by a different color, and the height of the rectangles symbolizes the level of each voice in the mix. The goal in this example is to even out the dynamics of the choir as a whole (the combined height of the rectangles). A compressor for each individual voice does the trick, and the resulting blend is very even for each of the chord repetitions (B). A bus compressor on the mix of the four voices retains the different internal balances in each chord, but controls the overall dynamics equally well. So if you are happy with the balances and contrast between the various elements in your mix it's better to use bus compression, while if you need to even out these balances it's better to use individual compressors.

with the mechanics of mixing through a bus compressor—there are benefits to be had. Just take a look at Figure 12.2, which shows you how a bus-compressed mix can appear to be more dynamic than a mix that uses compression on its individual components, while technically the two don't differ in dynamic range.

transient definition to withstand this, whether you'll eventually keep the compressor on your mix or not.

If you want to properly assess the effect of the compression you use, it's important that you don't just compare short sections. Putting compression on a mix changes all the dynamic proportions, including those between the intro, verses and choruses. If you only listen to the effect on the chorus, you won't know if it still has enough impact compared to the intro. This makes the musical effect of compression far more difficult to judge than that of EQ or reverb.

12.2 Choices and Settings

Just like the selection of glues on offer at the hardware store, the range of bus compressors is pretty overwhelming. But what makes a 'regular' compressor a good bus compressor? The simple answer is: as soon you put it on your mix bus, the transformation is complete. And if it sounds good, you don't have to think about it anymore. But there are reasons why some compressors are more favored for the entire mix, while others are mainly used for individual tracks.

The most important feature is that you can adjust the settings of the compressor accurately enough, since the role of a bus compressor can vary per mix. If you want to create a better blend between the vocals and the accompanying instruments in an acoustic song, the best choice will be a mild compressor that mainly follows the vocals. For example, a compressor with a low ratio, soft knee, slow attack and signal-dependent release can work well. But if you want the drums to make the entire mix move more, you'll need something with a response that's quick and tight enough to follow the drums. A compressor with a slightly harder knee and signal-independent attack and release times can suit this purpose. For most applications, it's important that you can set the ratio low enough, as high ratios will flatten the mix too much. And compressors that add a lot of distortion are often a bit too extreme to apply to the entire mix. In summary, for bus compression you need a compressor with versatile settings and a relatively clean sound, unless you have the luxury of choosing a specific compressor per project that works well for that particular mix.

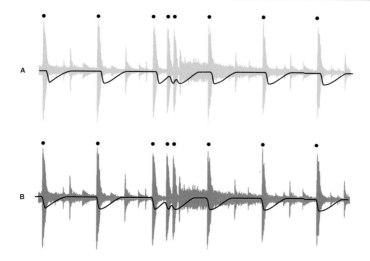

Figure 12.3 The timing setting that you choose strongly affects how the elements will be glued together in the mix. A slow attack (A) places more emphasis on the rhythm: the drum hits (black dots) will be largely allowed to pass, while the sounds that follow immediately after are attenuated. This can give the entire mix more punch and a somewhat dry feel. Conversely, a fast attack (B) already attenuates the mix during a drum hit. If you make the drums extra loud at the compressor's input, they will push the other instruments away when they play. As a result, the drums will seem louder than they actually are, while the rest of the mix will have less punch and more emphasis on the sustain—so it will feel less dry.

Stability

It seems like a luxury to have a lot of compression options for the mix bus, but this wealth of choices can also be a burden. Because the compressor affects all the mix balances you make, it's best to already have it turned on from the first moment you start working on your mix. This allows you to adjust your decisions to the compressor's response, so the overall mix will sound much more convincing than if you put a compressor on it during the final mixing stage. This also means that it's better to not change any settings on the compressor after you put it on your mix, because every time you do, you have to reconsider all the balance decisions you made before. This is why it's useful to spend some time on research and find out which compressor with which settings works well for a particular kind of mix. If you really start working on such a mix later, you can use your own presets and try not to change them throughout the course of the mix. You do need a certain amount

Multiband

Multiband compressors tend to sound very transparent when you first put them on your mix, but there's a catch. The power of mix bus compression as discussed in this chapter is in how it makes the various elements of a mix influence each other. If you divide the spectrum into separate frequency ranges, you'll lose an important part of this influence. For example, a bass drum can trigger the compressor, but since it only manipulates the lows in the mix, its influence on the rest of the mix will be very limited. With a regular compressor, you would hear the bass drum get so loud that it starts to push the rest of the mix away, but a multiband compressor only makes it softer. In some cases, this can be exactly what you're looking for, but it's not recommended to use multiband compression on your mix by default. Usually you're better off with compression on the individual elements, and maybe a sidechain filter in your bus compressor to make it respond less to specific frequency ranges.

of trust in your settings for this, but this kind of preparation also provides stability. If the compressor is a stable factor, you can focus all your attention on the actual mixing.

Sweet Spot

Often, there's a relatively limited zone in which bus compression actually sounds good. If you stay below it, the effect is negligible, but if you lay it on too thick, your mix will become lifeless. Therefore, the trick is to make the compressor function exactly in this zone. Whether this works mostly depends on your source material. If it's too dynamic, you'll need too much bus compression to glue it together, which will make the mix sound flat, smothered and forced. But if your source material is already (too) heavily compressed, there's little left for the bus compressor to do besides making the mix even flatter. So it's about the interaction between the compression on the individual channels and what you leave to the bus compressor. And sometimes it happens that bus compression actually improves the sound of your mix, but it flattens out the differences between the empty and full sections too much. In these cases, you can also automate the volume of the entire mix after running it through the

Stereo Link

Some bus compressors always work in stereo mode, which means that the left and right channels of the mix receive the exact same treatment in order to prevent shifts in the stereo image. So if your mix has a loud guitar part that's hard panned to the right, the compressor will push the entire mix down, instead of only the right half. But sometimes it's an option to turn the stereo link off, and there are even compressors that allow you to set the amount of stereo linkage. Especially in busy mixes, it can work to manipulate the two channels of the mix (completely or partly) independently of each other. This way, the compressor keeps the stereo image balanced in terms of energy, and it often brings out a bit more detail in the mix than if you link it. It's mainly a matter of trying out which setting sounds best, so you shouldn't blindly assume that stereo link is always necessary.

compressor (one decibel can already help a lot) to increase the differences again. Well-functioning bus compression can make a mix seem more musically dynamic (by creating more movement), while the dynamics on the meter are barely reduced.

Mastering

By applying bus compression, aren't you actually doing the mastering engineer's job, who has all the expensive gear (and years of experience) specifically aimed at manipulating the mix as a whole? It's important to distinguish between putting a compressor on the overall mix and mastering, because these two things aren't the same at all. Bus compression works best if you apply it to your mix from the very start. This way, it becomes a part of your mix, just like reverb on the vocals. Maybe the process is a bit harder to understand than adding reverb, but that doesn't mean you should pass it on to the mastering phase. Bus compression affects the mix balance, and that's something you want to hear while you're still mixing. However, it's recommended to distinguish between compression and limiting. The latter does virtually nothing for the cohesion of your mix, and only serves to gain absolute loudness. Therefore, it's pretty pointless to mix through a limiter, and on top of this, limiting (unlike correctly applied bus compression) restricts your mastering engineer's options.

12.3 Multibus Setups: Painting on a Colored Canvas

As mentioned before, bus compression isn't used to reduce the dynamics of your mix, but to increase the cohesion between instruments. However, since reducing dynamics is the core business of a compressor, it's impossible to create more cohesion without also flattening the dynamics of your mix. Sometimes this is a problem, especially if you're leaning heavily on compression to mold things together. To counteract the sonic shrinkage that can occur as a result, you can spread the mix across multiple mix buses, each with its own bus compressor. This way, many hands will make light work, instead of one compressor having to deal with everything at once. The fun thing about this system is that you can choose a number of different compressors, each of which enhances a different musical aspect and produces a different kind of blend. You can see this approach as painting on a colored canvas: your canvas influences the place that's given to a particular element in the painting. The famous mixing engineer Michael Brauer has turned this into his trademark, but many others have also used versions of this method for years.

Mix buses that are designed to give sounds a specific place can make it a bit easier for you to set up your mix. You can see it as a way to automatically apply the correct incidence of light while you paint. If you really want to highlight a particular element, you can place it in a lighter zone than when it plays a supporting role. If an instrument contributes to the rhythmic foundation, you send it to a bus that emphasizes the rhythm. A lead instrument goes into a bus that brings the sound out to the front. An instrument that's tonally important goes into a bus that

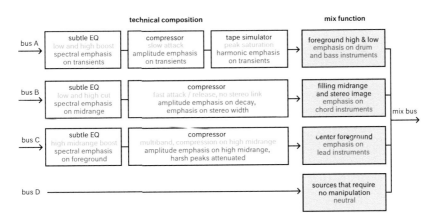

Figure 12.4 An example of a setup with differently colored mix buses, which reserve specific places in the mix for the sounds you send through.

emphasizes the sustaining character of the sound. This system makes it easier to balance things quickly, because you have more zones to play with than a white canvas can offer you. On a white canvas, you would have to reinvent the wheel for every element you add in order to position it in the whole. Predesigned mix buses provide a shortcut for this, and of course you can still fine-tune the sound as much as you want after you've outlined the general shape of the mix.

It might sound crazy to mix with a setup that instantly affects a variety of parameters, but the interaction between your setup and the music makes it easier for you to make certain choices, because they will immediately sound more convincing. It's as if your setup helps you to hear when your balance is on the right track. Developing such a colored canvas takes time, and the effect of it is more subtle than a lot of beginners think. Caution is advised though, as a colored canvas can also adversely affect your mix, for example by flattening the dynamics too much. In that sense, it's no different than using a single mix bus compressor. To keep an eye on the influence of the setup, in case of doubt I sometimes switch to a fourth bus. This bus sends the sound into the mix without any manipulation, so I can always check if the tools don't get in the way of the music.

Chapter 13

Templates
Working Faster Means a Better Mix

This chapter is about making presets for your mix: basic settings that you can almost always rely on. This might sound like a recipe for making the same mix over and over again, but I would argue that the opposite is true. A familiar starting point gives you the confidence and ease required to make decisions quickly, leaving you more time to get to the core of the music. Hardware can speed up this process a bit, because it settles into your muscle memory. You've probably seen how experienced engineers can really fly across the mixing desk. But it can also be a burden if you have to switch between projects a lot and restore the settings every time. The last part of this chapter discusses how you can address this problem.

13.1 Time Is Money

The more often you hear a piece of music, the more you get used to it, and the harder it becomes to still judge it objectively. Once you start mixing, you want to utilize every second you spend listening to the music as best you can, especially during the first stages of the process, when your ears are still fresh. I'm pretty strict when it comes to this: after I've heard the rough balance once, while I'm preparing the mix (sorting and naming the tracks, setting up the routing), I don't listen to anything anymore. I want the first few times that I play a song to be as effective as possible. If these initial plays give me an idea for a mixing approach, I should be able to immediately test this idea. This way, I get to know a lot about the material. During this testing phase, finding out what doesn't work is almost as educational as finding out what does, which is why it's so important to try things out. If you don't do this, you won't learn anything. Ideas that seem amazing in your head can be very disappointing in practice.

The first phase of a mix resembles the rough sketch a painter makes on the canvas. The details can be filled in later: first, the main decisions about the shape of the composition should be tested. Translating this process to your mix means that you should be able to quickly control the most important mix parameters. Besides volume balance, these are timbre, space and dynamics. A mix template can help you to quickly make these parameters manageable. Such a template consists of a number of 'proven formulas' for routing, EQ, compression and effects that you can use as a starting point.

In a way, an analog mixing console can also be seen as a template: all the choices in terms of which basic EQ and compressor you want to use have already been made (sometimes the settings are still there from the previous mix), plus usually you already have effect sends connected to external reverbs, and sometimes processors plugged in as inserts on a subgroup. In a DAW, this is very different, as you completely start from scratch if you don't develop templates yourself. These templates are anything but a way to keep repeating the same trick. On the contrary: they can tell you right away that the method that usually works doesn't work this time around, and you'll have to think in another direction.

Organization

It's not recommended to put default EQ, compression and reverb on all the channels from the get-go, through a system of presets. A template should be about the big picture, not about the individual details that are different for every mix. Take vocals for example: after a hundred or so mixes, you come to the conclusion that—though you might need different EQ corrections for each voice—you pretty much always add some high mids and use the same plugin to add some compression. In this case, you can already prepare this high mid boost and compression in a template. The nice thing about this kind of template is that you can immediately dive into the things that make the voice unique if the general approach works. And if it doesn't work, the resulting sound will usually give you an idea about what will.

Why do you often end up with the same kind of manipulations in mixing? Aren't all mixes different? True, but conceptually, it always comes down to the same thing: enhancing musical parameters such as rhythm, tone and dynamics, and solving problems such as lack of definition or messy dynamics. And while it differs per mix which instruments provide these parameters and to what extent your interventions are needed, the formulas you develop for these interventions are pretty widely applicable. You can see these formulas as your unique toolbox, which can be used in various degrees in all your mixes. Therefore, the artistic part of mixing is less about your toolbox (although you might come up with unique solutions, of course) than about knowing when and how to use it. This is why many famous mixing engineers are very open about their setup and sometimes even publish their DAW templates, as they are the only ones who can really work effectively with these systems.

13.2 Building Templates

The most important function of a mix template is to organize the mix components according to a fixed pattern. If you want to work quickly, the main thing is to avoid spending a lot of time figuring out where a particular sound is coming from. This is not so easy these days, with sessions that can easily have a hundred source channels. Therefore, the first step I always take is to divide these channels into a number of groups (sixteen at most, a number most fader controllers can handle well), which I put together based on instrument groups (so all drum microphones end up in a drums group, all overdubbed percussion parts in a percussion group, a string quartet made up of close-miked instruments and a main microphone pair also ends up in one group, and so on). With these sixteen faders, I can quickly outline all the mix balances I want to try out, which makes experimentation easy.

During this outlining stage, I usually hear a couple of things I want to change in some instruments. To quickly fix these, I use a preset chain that can apply a rough approximation of the adjustment I have in mind. Examples are presets for making things bigger, wider, warmer or more aggressive. These settings can still be refined later, but initially I'm only drawing a rough sketch (see Figure 13.1). Sometimes I also merge several instrument groups together into a single 'supergroup,' if they share a particular musical function that I want to emphasize. An example of this is a rhythm created by a combination of drums, handclaps and keyboards. If I manipulate these instruments as one group with EQ, compression or whatever other effects I want to use, the rhythm will automatically start to feel like even more of a whole.

One Turn of the Knob

When setting up an initial mix, you usually only need your faders, pan pots and the routing discussed above. But sometimes you can immediately hear that certain sounds aren't working yet in their current form. In these cases, it's important to quickly force them into a mold that does work, because otherwise one flawed sound can hold back the progress of your entire mix. This process can be facilitated by putting the same EQ and compressor/gate on each channel (which means you're actually mimicking an analog mixing console).

But there's an even easier way to do this: the aux-sends can be used to connect all kinds of parallel effects. The main advantage of this is that you can quickly try out a couple of basic sonic characters to give the sounds a certain perspective. These characters can be used for multiple sources, and you don't have to change their settings every time you want to use them. One turn of the knob is enough to immediately hear if it works on a particular source. If you categorize these aux-sends according

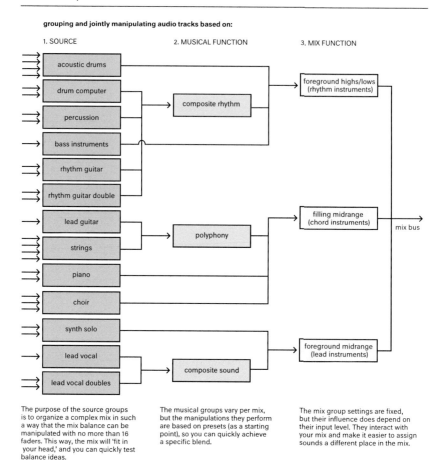

Figure 13.1 Three levels for grouping and manipulating audio.

to a 'best of' selection of your own effect settings, you can set up an initial rough mix in no time. And as soon as you want to try something else, you can always replace effects or change their settings.

Besides for spatial effects, the aux-send approach is also highly suitable for compression (see Figure 13.2). Parallel compression that's connected this way can keep sounds from disappearing in the mix: it guards the lower limit, whereas insert compression keeps the peaks in check (see section 5.3 for more explanation on this topic). Disappearing sounds are usually the biggest problem during the sketch phase of the mix, so it's very useful to have a couple of parallel compressors ready when you start mixing. And if you ever need more compression than 50 percent, you can connect the aux pre-fader and then turn down the source sound in the

Gain Structure

A template can only work if its signal levels are calibrated. Calibration is important, because you don't want sounds to make huge level jumps when you send them to another bus (which can easily be adjusted by running an equally loud test tone through all the buses). But on top of this, mixing chains made up of compressors and/or distortion-adding devices also have a sweet spot. Ideally, your template is designed in such a way that the signal levels in the chains end up exactly in those sweet spots (with average mix balances, at least). This way, your balance choices will have the biggest effect: besides changing the level they will also change the sound and perspective. Test tones are less suitable for adjusting compressor thresholds or amounts of overdrive, as they're too static for this. It's better to determine these settings empirically during mixing.

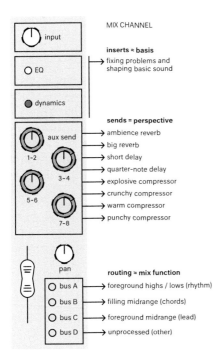

Figure 13.2 This is what your template's channel strip could look like. The EQ and dynamics correct the source sound if necessary, the auxes add a certain color to it, and the routing determines where it ends up in the mix.

mix, independently of the compressed sound. So all in all, it's a pretty flexible approach, plus it makes very efficient use of the available equipment (or processing power). However, keep in mind that for parallel compression, good delay compensation (and/or hardware that doesn't cause big phase shifts) is essential, otherwise comb filtering will occur.

Stocking the Pantry

To develop your templates, it's important to keep your eyes open for techniques that affect the sound in a way that could come in handy again in the future. For example, a chorus effect that you created might simply be a chorus effect to you—and then you slowly forget about it again. But you can also listen to what it really does: maybe it stretches a sound beyond the width of the speakers. The latter is a solution you might need much more often. Over the years, I've come to look at equipment this way: not at what it is, but rather at what it does. For example, I found myself dubbing certain combinations of EQ and compression 'warmer-upper,' 'cutting torch,' 'magnifier,' 'restrainer' and 'focus puller.'

By giving technology a personality, you can make it easier to apply. Over time, you can develop characters for all frequently used manipulations: rounding off transients, widening the stereo image, applying focus and density in the midrange, emphasizing the rhythm or the space, enhancing tonality, adding air, and so on. These prefab solutions aren't a standard part of the template, like routing and bus processing. Instead, you keep them stored in the pantry, ready to be used whenever you need them. Your template can be as complex as you want it to be. You can turn it into a real masterpiece, made up of built-in sidechain constructions, thirty different send effects, bussing, tracks to record stems with, and so on. But keep in mind that your template should work as a catalyst, not as a complication. It should be as universally applicable as possible for the work you want to do with it.

Tilt EQ: A One-Knob Equalizer

If the problems in your mix aren't too extreme, most corrections can easily be made with very broad EQ curves. Often a bit more or less bass or treble is all it takes to get an instrument where it should be. A device that can be very useful for this purpose is a tilt filter, which consists of a high-shelf and a low-shelf filter around the same center frequency. These two shelf filters work inversely in terms of amplification and attenuation, so more high end automatically means less low end, and vice versa. Because a tilt filter can be controlled with a single knob, it's the perfect candidate for inclusion in your template. This will make EQ just as quick to control as your faders, pan pots and parallel compression, which is ideal during the sketch phase of your mix.

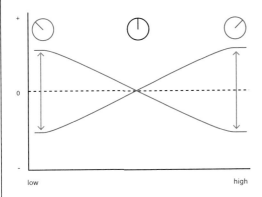

Figure 13.3 With a tilt equalizer, you can tilt the entire frequency spectrum with a single knob: downward (blue) or upward (red).

13.3 Integrating Equipment in the 21st Century

Hardware can stimulate your creativity and result in unique sounds. But the speed and flexibility of a computer isn't so bad either, especially the possibility of instantly trying out any idea that comes into your head. With just a couple of mouse clicks, you can duplicate a channel, create a complex routing or choose any effect you can think of. Everything is saved in your project, and with a laptop and a set of headphones you can work on it wherever you are in the world. The possibilities this offers are nearly endless, so the thought of restricting this workflow by using all kinds of hardware can be pretty frightening.

Recall

The main problem with using external equipment is retrieving previous settings if you still want to change things a week later. Especially if you're mixing through an analog mixing console, it's a time-consuming task to put all those little knobs back in the exact same position. Even with expensive mixing desks with total recall systems (which partially automate this process) it takes time and effort, and it still won't sound exactly the same as the original mix. In the old days, studio assistants would write down the settings of each device on sheets after every production, but fortunately, thanks to digital photography, those days are over. With a good camera, you can capture the settings of multiple devices in a single photo. If you then copy these pictures to your tablet, you can hold them next to the device when you're restoring the settings.

If you've ever used this method, you'll immediately understand importance of a clear legend and easily readable knobs. Stepped attenuators instead of potentiometers are even better—but expensive. On top of this, the market for digitally controlled analog devices is still growing.

Freeze

Photography might make it quicker to save your settings, but restoring them still takes a lot of time. It's for this reason that I always divide my mixes into separate instrument and effect groups, which I record onto empty tracks (stems) at the end of a mixing day. This way, I only need to restore the devices of the stems I still want to adjust later, and of the mix bus where the stems are routed through. The effects on individual tracks can be saved with the freeze function (which many DAWs have on board). This is nothing more than a recording of the output of a particular track, including all internal and external effects, which is used in the project as a replacement for the original audio and effects. This is a very useful feature, because after you've frozen your external effects, you can use the newly available processing capacity for other instruments.

Another advantage of recording stems is that it makes your project portable, so you can also work on it somewhere else. In my own setup, I usually only need to recall the mix bus chain, while all the other mix adjustments can simply be made on the recorded stems. Most of the time, a little fine-tuning is all you need after the first mix version has been recorded. For example, with an EQ plugin, you can easily cut the lows of your analog-processed drum bus a bit more, without having to restore the settings of the entire drum chain to redo the whole thing. This working method also allows for automation after the use of analog EQ and compression. This way, you prevent the compressor from working less hard when you turn the vocals down.

Setups with analog summing mixers make this scenario much more difficult, unless you accept that you can't record the manipulated stems (not a lot of summing mixers have direct outputs), which means you have to completely restore your equipment for every single mix revision. If you then also want to use post-compression automation, you'll even need a digitally automated summing mixer.

For me personally, the convenience of a setup that allows me to record all the manipulations made with external equipment onto new tracks outweighs the sonic advantages of analog summing. But since I still want

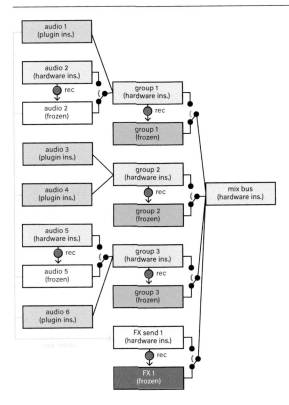

Figure 13.4 A mix with hardware inserts on three levels. If you record the individual
tracks, groups and send effects separately, you often only need to
restore the settings of the hardware on the mix bus if you still want to
adjust things later. And otherwise you only restore the equipment of
the groups and tracks you want to adjust. Of course, this recording will
have to take place post-fader, and therefore including automation. And
if you want to retrieve a send effect, you'll have to run the recorded
tracks and groups through it again.

to reap some of the benefits of using analog circuitry on the whole mix,
I do send these recorded channels through a multibus compression setup
(as discussed in Chapter 12) to an analog mixer. However, the settings
on this mixer are always the same, so the system still has 'total recall.'
Of course, your preference can be different from mine, for instance if you
like the sound and ergonomics of a specific analog mixer.

To Patch or Not to Patch . . .

Using hardware seems like fun, until you end up sorting through a tangle
of patch cables for minutes every time you want to use a device.

Levels

All analog equipment has a range in which it functions best. If the signal levels are too low at the input, the noise floor will get too close to your signal, while high levels can lead to hard distortion. Therefore, it's important to provide each analog device with the

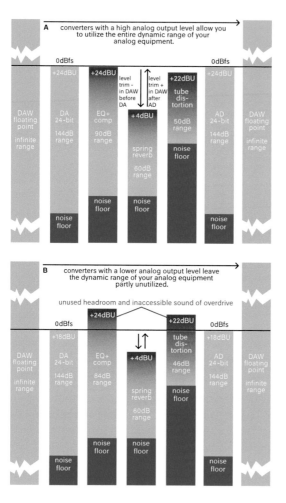

Figure 13.5 Good gain structure provides the best possible signal-to-noise ratio, and total control over the amount of coloration added by your analog devices. This is achieved by raising or lowering the levels you send to these devices.

right signal level. This sounds like a lot of hassle, but it can be easily controlled from within your DAW, which means you can also store these levels as a preset. High-quality 24-bit DA/AD conversion allows you to make signals weaker at the output or input, without running into noise problems. Using digital trim plugins—which you place before and after the hardware inserts in your DAW—you can easily set the ideal analog level for every device in the chain. This is very useful, because a tape echo will start distorting at a lower signal level than a mastering equalizer. If your converters can send out a high analog signal level, you can access the entire usable operating range of your analog equipment without ever making your converters clip (Figure 13.5A).

If the maximum usable level of your analog equipment is higher than that of your converters, or if you don't set your trim plugins properly, you're wasting dynamic range. This also means that for some devices the input level will be too low to make them distort (slightly), so you won't be able to use the edge these devices can add to the sound (Figure 13.5B).

Fortunately you can make your life a lot easier, simply by having enough converter channels available. You can then connect your equipment directly to the converters, instead of through a patchbay. This way, it's just as quick and easy to choose a hardware insert in your DAW as a plugin. Only if you want to create multiple chains of devices is this approach not so efficient, since every device requires its own DA/AD conversion. This is why I create fixed hardware chains (for instance an equalizer and compressor that work together well) that I connect to the converters as a whole.

Devices that you often combine in different sequences can also be connected through an analog switcher (optionally with a digital memory for the created routing), a setup that allows you to create and recall your routings with a push of a button. These kinds of handy tricks might seem like an unnecessary luxury, but saving time is literally the reason why some studios use plugins when the original hardware is in a rack two steps away.

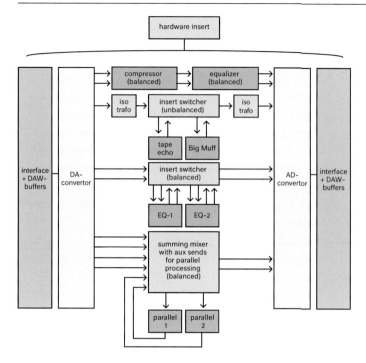

Figure 13.6 With this setup, you don't need a patchbay: all equipment is permanently connected to converter channels. Unbalanced devices can be connected with isolation transformers to keep unbalanced cables as short as possible, and to prevent hum. The routing can be controlled from your DAW mixer and by using analog tools such as the insert switcher and mixer.

Delay Compensation

With any setup that combines hardware and software, the success depends on delay compensation. AD/DA conversion, DAW buffers and plugins all cause delay, which must be compensated for with military precision to prevent phase problems. Although most DAWs automatically correct any delay that's known to the application, you won't know for sure that your system is fully synchronized unless you test it. Some DAWs offer the option of measuring and automatically compensating the delay caused by hardware inserts. They do this by sending a pulse through the chain, but unfortunately this doesn't always produce a correct result.

Especially with effects that add a lot of distortion (and of course reverbs), the delay is hard to determine. While measuring, I often put these effects in bypass mode. After the measurement, the easiest way to verify if your entire system is synchronized is by placing a test file with a number of pulses in each audio track. After playing these tracks through your mixer and all of the internal and external effects you use, the pulses of all tracks combined should be perfectly aligned in the mix that comes out on the other side (except for some phase shifts caused by EQ, and decay caused by reverb and delay). The only problems that can still arise have to do with polarity: some hardware switches the plus and minus of the signal, resulting in a pulse that seems synchronous, but when you use the effect in parallel, it still causes attenuation in the mix.

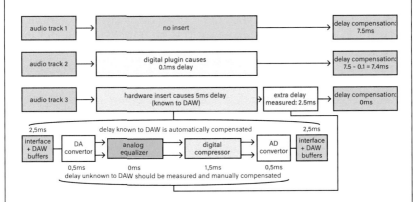

Figure 13.7 Delay compensation in a digital mixer with plugin inserts and hardware inserts. The amount of delay that each track gets is measured in such a way that, at the end of the chain, all tracks are aligned with the track that's delayed the most. Keep in mind that automatic delay compensation can be wrong: the delay caused by digital hardware and separate AD/DA converters isn't known to the DAW. Therefore, this delay will have to be measured and manually compensated for.

Chapter 14

Preparing for Mastering
A Good Relationship

Mastering is the last creative step in the music production process. It's a task you preferably want to entrust to a specialized engineer. Someone who listens to your work with an unbiased ear, in a room with ruthlessly good acoustics and monitoring, and who will make the final corrections in the frequency balance, dynamics and stereo image. But before you upload your mix to a mastering engineer, it's useful to consider what you can expect from him or her.

In short, the goal of mastering is to optimize the mix for the intended audience. This target audience is often very broad, in which case mastering is aimed at creating a sound that works on the radio, on earbuds and on a PA. But sometimes the target audience is more specific, like 'classical music aficionados' or 'people who attend raves.' A good mastering engineer takes into account how people experience music in a certain context. In every genre, the way we experience music is tied to different things, so it varies per mix which aspects are important. There are no such things as optimal dynamics or a universal frequency balance. Punk rock, for example, has a much shriller sound than R&B, and for good reason. The goal of mastering is to capture the musical experience as well as possible on the intended medium (vinyl, radio, streaming services, and so on). This chapter discusses the mastering process, focusing on the question of how you can get the most out of it as a mixer.

14.1 Communication

The unbiased ear of mastering engineers is their greatest strength, but it's also potentially the biggest bottleneck in the mastering process. Since you can't expect someone without prior knowledge to immediately understand all the nuances of your music and production approach, it's always a good idea to attend a mastering session. This way, you will develop a relationship with the engineer in question, which could prevent misunderstandings in the future. Understanding why a mastering engineer does things a certain way will help you to better assess the master.

Otherwise, chances are that you end up listening to the master on the same system(s) on which you listened to the mix—and where you've grown accustomed to the sound of the mix. You're bound to hear every change as something that doesn't belong there: it's conflicting with your own memory. This is less likely to happen if the mastering engineer explains (and also demonstrates in the mastering studio) why the highs had to be attenuated, for example. Back home, you'll pay more attention to the difference in depth this produces, instead of only noticing that the master is less bright than the mix. It also works the other way around: mastering engineers will make other choices if they know what you care about. If the guitars are the most important element in your mix, the mastering engineer won't try every trick in the book to make the vocals come to the front.

Feedback

Of course it's possible to hit the bull's-eye with the first attempt, but with more extensive projects (like an album), it's not an unnecessary luxury to reflect on the work of the mastering engineer (which will also benefit the engineer, by the way). During a day of mastering, there's hardly any time to listen to an album in full. At this point in the process, even the mastering engineer will have trouble hearing the music over all the

---

Reference Tracks

Reference tracks can help to communicate your preferences to the mastering engineer, but they can also draw attention away from the essence of your own production. Try to use your references as an indication of the context you want your production to relate to. The mastering engineer will then aim for a similar impression in terms of perspective and impact. Or at least to a certain extent, because there's a point where you have to realize that every production sounds different, and for a reason. All your production choices (such as arrangement, microphone selection and mix) affect the final result in some way, and mastering is only a part of this whole. Good mastering enhances your production choices, while bad mastering tries to turn the mix into something it's not. This is why it's better for reference tracks to play a role earlier in the process than during mastering, though they can be useful as a starting point for a conversation.

technical mastering choices. Therefore, it's a good idea to listen to the master again after a couple of days and in different settings, and then make minor adjustments. Make sure to discuss with the mastering engineer in advance if such a feedback round is included in the price, or if every revision costs extra money.

When you judge a master, try not to listen to it as a mastering engineer. Instead of thinking about the amount of low end and the kind of compression, it's much more important for you to feel if the music comes across. It's often easier to do this if you don't listen with full attention, which is why the car is a reference for a lot of people. Having friends over while the music is playing in the background can also do wonders. This half-attentive method is great if you only want to pick out the few points that really need more work, and entrust the other choices to the mastering engineer's expertise. If you plan to compare the master with the mix or other reference material, it's important to take the time for this (listen

---

Test Masters

If you don't know the mastering engineer yet, naturally you're wondering if he or she can take your work to a higher level. A relatively accessible way to find this out is by first ordering masters of one or two tracks. Of course, this will cost you some money, but it will give you a much more accurate impression than a free test master of a minute of music. The bigger mastering studios never do free tests, and that's not out of snobbery. The serious mastering of a single track—or even just a fragment of a track—is a lot of work. In the case of an album, the first track can easily take two hours, while the remaining ten tracks might be finished during the rest of the day. The search for the approach that fits the identity of the music the best is by far the most work. Once the approach is clear, it's relatively easy to continue with the work (which is why mastering a separate track is often a bit more expensive than a track on an album). So if you ask for a free test master, in many cases your wish won't be granted. But if you do manage to get one, the approach will often be much less thorough, resulting in a quick demo that's not very representative. On top of this, there is no room for communication with the mastering engineer, so there can't be any mutual understanding on which direction to take, and the entire process becomes a bit of a gamble.

to the master as a whole, not in bits and pieces) and compensate for the differences in loudness, for example by putting all the files in iTunes and turning the Sound Check option on. If the production involves a band with several members, listening separately can also be a way for the main points of criticism to emerge, as these will be mentioned unanimously.

14.2 What Is a Good Mix?

Where does the work of the mix engineer end and that of the mastering engineer begin? In practice, this boundary always lies somewhere else, and you can roughly say that the higher the quality of the production and mix, the smaller the difference that mastering makes. Not that mastering is less important for top-quality productions, but just don't expect dramatic differences between the mix and the master. Which is good, because you can assume that the artist, producer, mixer and record company know how they want things to sound if they spend so much time, love and money on a project. Producers and mixers who reassure their clients with the statement that the problems they hear will be solved in mastering don't take their clients seriously. The goal is to deliver a mix made up of choices that the client supports one hundred percent. Whether this mix can get even better is then a question for the mastering engineer to answer, but he or she must be able to build on a foundation that's approved by all those involved. A mastering engineer can't perform magic: if it's not in the mix, it can't be created through mastering.

A good mix is the result of making clear choices that are inspired by the music. During mixing, you shouldn't worry too much about whether there's too much bass or whether the vocals are too loud. It's about creating something that's bursting with the power of the music, something that gives you a good feeling, right there and then. That's why it's so nice to know that there will still be a mastering engineer working on your mix, someone who watches your back and who will make sure that the feeling you get from your mix will also translate well to the rest of the world. This makes it easier for you to follow this feeling when you mix.

This way, mastering can give a mixer more creative freedom. Of course, this doesn't mean you can do whatever you want, as it's still extremely important that the various elements are correctly balanced in terms of volume. Especially the proportions in the midrange are crucial, so it's a good idea to check these on a small speaker. If the volume balances work, the mastering engineer can still use EQ to solve a lot of problems in the frequency balance or give the overall sound a certain focus, without the mix falling apart.

Your production has to speak for itself, so it helps if you don't compare it to the rest of the world too much. Try to create a mix that you yourself are impressed with, without emulating the sound—and particularly the loudness—of reference tracks. This requires trust in your own abilities and judgment: mixing is not for doubters! However, what you can do is listen to the balance choices and atmosphere in the reference tracks. The best way to do this is by playing them at the same loudness as your own mix. This will prevent you from making a mix that might be loud enough, but that has balances and a timbre that don't match the music at all.

If your mix is well-balanced, and feels energetic when you turn it up in your own studio, the mastering engineer can take care of all the extra loudness you might need. Here are some more general recommendations:

> - Don't put off making choices because you're afraid they might complicate the mastering engineer's job. Create a mix that's as close as possible to what you have in mind, except for the absolute loudness.
> - Check your mix doesn't fall apart in mono.
> - Check the proportions between instruments on different types of speakers.
> - Check your frequency balance and dynamics at different monitoring levels.
> - It's better for your mix to have too much contrast and impact than too little. Important things should really pop out of the mix.

14.3 Mix Bus Processing

The question whether you should submit your mix with or without mix bus compression has already been answered in Chapter 12, but it's important to distinguish between 'real' mix bus compression (which you put on the mix while mixing) and 'pseudo mastering' (compression that you put on your mix afterwards). The second category is usually best avoided, while the first one will become an indispensable part of your mix. If you submit your mix without it, the mastering engineer will have to precisely re-create its effect, which is virtually impossible without the exact same compressor.

In other words, there's nothing wrong with mix bus compression, and as long as you use it wisely, it won't limit the mastering engineer's options in any way. In fact, if you're looking for a certain cohesion and density in the sound, this is much easier to achieve with mix bus compression than with mastering alone. The mix balance you make while hearing the compressor's effect will automatically be more compression-resistant than the balance you would make if you didn't hear how it sounds with compression.

Tech Specs

Although many mastering studios still accept source material on analog tape, nowadays almost everything is submitted digitally. The best format for this is an interleaved WAV or AIFF file, at a resolution of 24 or 32 bits floating point. It's pointless to use a higher sample rate than that of your source material if you're mixing in the box. Always submit the mix at the sample rate your project is set at, as each conversion is detrimental to the sound quality. Besides this, for in-the-box projects, it's a good idea to work at a high sample rate (88.2 kHz or higher) from the start, as this will save you a lot of sample rate conversion in the plugins you use. However, if you work on an analog mixing desk or a summing mixer, it's useful to record the final mix at a higher sample rate than the source material, and use the best AD converter you have for this. Make sure the peak levels of the mix are roughly between −10 and −3 dBFS, and don't use any peak limiting (see section 14.3 on mix bus processing).

The files can be submitted on a USB stick, but it's wise to upload them as well, so the mastering engineer can check the files for problems before the session starts. It's smart to compress files into zip files, as these can only be uncompressed if the data has arrived intact. Carefully listen to the mixes before you send them: it won't be the first time that a channel has accidentally been turned off in the mix bounce. Use clear naming with version numbers, so there can never be any confusion about which file to use if there are more versions of a mix. And if you can't attend the session in person, inform the mastering engineer of your wishes for transitions between songs (crossfades, specific breaks, fadeouts, extra track markers, and so on).

Mastering engineers generally don't get very excited when you use a mix bus compressor in multiband mode, or when there's an exciter or limiter on your entire mix. Multiband flattens the frequency spectrum, resulting in a mix balance that sounds very controlled—but usually also weak. Exciters make everything more crisp and bright, but they also eat up all the depth and the distinction between parts in the high end. The definition they produce is 'fake,' in a way. There's nothing wrong with this if it involves individual parts, but if you need it to save your mix, it's usually a sign that there's something wrong 'under the hood.' The same can be said for limiters, as they generally only make the mix weaker, and they eat up definition. Unlike compressors, they do little to improve your sense of balance and energy—it's all about absolute loudness.

What a limiter can do is tell you if the structure of your mix is solid enough. If you can't seem to make your mix as loud as a good-sounding reference track through limiting, this usually points to problems with the mix balance, EQ (masking) or compression you use. In this regard, a test version with limiting can provide some insight into the power and efficiency of your mix, both for yourself and for those clients who get nervous if the mix isn't loud enough. I always check these things by calibrating my loudest monitor level (I know how impressive music is supposed to sound at that level) and by looking at my loudness meter. If the mix looks loud on the meter but doesn't sound loud, I know I'm still wasting too much energy. You can read more on loudness measurement in section 14.5.

Vinyl

Although the mastering engineer is usually more than able to monitor the specific requirements involved in a vinyl release, there are a number of things you can already take into account in the mix. This way, you'll avoid having to make major compromises in mastering. Vinyl has two main limitations in terms of sound: it can't handle large concentrations of high frequencies, and it can't deal with large phase differences between left and right in the low end. The mastering engineer can curb both of these excesses, but this is not without a price: if you need to make the lows more mono due to a single out-of-phase bass synthesizer, you'll also lose the low-frequency width in the other parts. Therefore, it's advisable to keep an eye on your correlation meter during mixing (see section 11.4), and to make sure there's not too much low-range energy out of phase. The same goes for high frequencies that stick out above the mix: de-essing the problematic part usually sounds better than de-essing the entire mix.

This doesn't mean that vinyl requires a completely different mixing approach, as the excesses mentioned above often don't sound very nice on a digital medium either. The only difference is that vinyl starts to protest physically. In this respect, it's also not a good idea to use too much distortion (think of tape, tube or transformer saturation) on the overall mix. Vinyl will still add some of its own distortion to this, so before you know it, it's too much of a good thing. The mastering engineer can make a precise estimate of the amount of coloration that will still work on vinyl. One last aspect to keep in mind is the available time per side: a 33 rpm LP sounds best if there's less than twenty minutes of music per side, even more so if the music has a high sonic density. It's important to evaluate how an album comes out of the pressing process before you start making thousands of them. Therefore you should always request a test pressing and check it at home.

Despite all its limitations, vinyl is a wonderful medium. Not necessarily because of the sound, but because you can count on it that a vinyl listener will hear your work in full, on an above-average sound system. It's a medium that forces people to focus on the music, that allows for longer dramatic arcs, and that's unburdened by feverish loudness comparisons with other productions. For me personally, it's nice to think that people will take the time to listen to my work. It encourages me to make more daring choices, because I don't have to be so afraid that listeners will tune out if the music doesn't get to the point within thirty seconds.

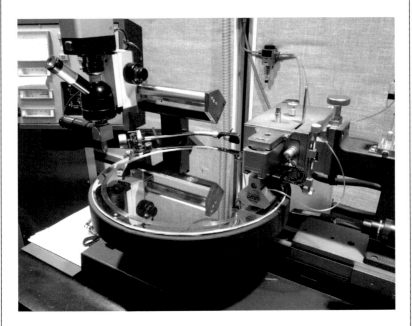

Figure 14.1 Cutting a vinyl master

14.4 Stems

Not everyone is an experienced mix engineer, and therefore not everyone will submit equally useful mixes for mastering. In case of subpar mixes, mastering engineers can give feedback and ask for a number of adjustments—or work like hell trying to find a remedy for the problems. But usually the best thing to do in these cases is to master from a number of stems. These are stereo groups of the most defining parts in a mix. A common division is, for example: drums, bass, chord instruments, vocals and effects. If you play these groups together with the faders at zero, they should sound exactly the same as the stereo mix. However, if you've used bus compression on the mix, it won't be so easy to pull this off, but it's not impossible (see Figure 14.2). The stems give the mastering engineer more precise control over the way the main elements work together. The trick is to keep the number of stems limited, because before you know it, the mastering engineer has to reconsider all the mix decisions and can't get around to actual mastering anymore. In any case, stem mastering

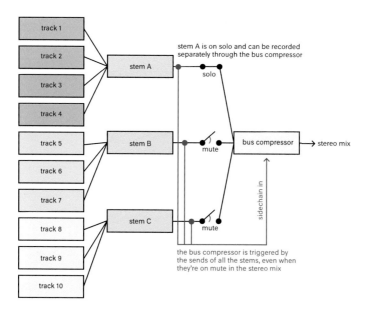

Figure 14.2 The use of bus or sidechain compression can make it pretty difficult to export stems. The trick is to make the compressor react as if all the tracks in the mix are on. You can do this by sending all the tracks (or a recorded mix without compression) to the external sidechain input of the compressor. This way, it will always react the same way, even though most channels are on mute.

costs more time—and time is money. Nevertheless, the cost–benefit ratio can be very good: stem mastering can make the difference between a mix that doesn't work and a professional-sounding master.

If there's no need for stems (the mastering engineer will tell you), don't use them. A mastering engineer's strong suit is translating production choices that are already made, not reconsidering them. If certain problems are better solved in the mix, the mastering engineer can make a specific request. With in-the-box mixes, it's even possible to export a modified mix from your laptop, if you bring it with you to the mastering studio. If you don't have this option, it's a good idea to export a few alternative versions of the mix before the start of the mastering session, with the vocals 1 dB louder and 1 dB weaker, and the same thing with the drums, for instance. In practice, I end up using these alternative versions in ten percent of cases. Usually the mix balances are good the way they are, but in the context of an album you sometimes need to make slightly different choices. Especially the placement of the voice determines how a song relates to the songs surrounding it. And the loudness of the drums determines how well the mix can withstand interventions with compression or limiting.

Whatever the exact method is, if you want to get the most out of your mastering session, a good understanding with the mastering engineer is vital. Even big names like Bob Ludwig or Greg Calbi don't claim to have all the answers. For them, it's also a matter of finding a way to make their clients' ideas work. And those ideas are difficult to distill from an email if you haven't met each other yet. Therefore, I advise everyone to be present at mastering sessions and ask a lot of questions, because this is the best way for a mutual understanding to develop naturally. On top of this, it's very educational: my own mixes are still improving as a result of the feedback I get from mastering engineers.

14.5 Consequences of Loudness Normalization

I like dynamic music, but if there's anything I find annoying, it's hearing my own work on a playlist at a party, but at a lower level than the previous track. It's a bummer, and even though you get used to it after a minute, it's definitely not a good first impression. Therefore, in mastering, I make sure to use all the loudness I can gain with a minimal compromise on dynamics, and I used to go even further if clients specifically asked for it. These days, at most house parties the music is coming from a playlist in Spotify, iTunes, Tidal, or even YouTube. And since these streaming services started using a normalization function, the loudness laws have changed completely: loudly mastered productions are relentlessly turned down, often even below the level of dynamically mastered productions.

A similar system is already in use in broadcasting, and a lot of professional DJ software and media players also have built-in normalization functions. It's only a matter of time before normalization is universally introduced, and virtually all digital music distribution is aligned in terms of loudness. Only for CD and vinyl, there's no standard. But since CD changers and homemade compilation CDs have gone out of style, there's no real need for an absolute loudness norm here either. Now it's up to you—and your clients—to properly judge the loudness of your productions: if you do this without normalization, the 'loudness cycle' will continue to seduce you, but you're in for a rude awakening the next time you go to a house party . . .

How Do You Measure Loudness?

Normalization is only interesting if it uses the right quantity. In the old situation, the peak level of the audio was all that mattered. There was a specific maximum level that you couldn't exceed: 0 dBFS, but nothing stopped you from using limiting and clipping to beef up the density—and therefore the loudness—below that peak level as much as you deemed responsible. So if you really want to say something about loudness, you'll have to look at the density (the average or RMS level) of the sound. But if you've ever compared unweighted RMS measurements of different productions, you'll know that they don't necessarily reflect the perceived loudness. It says a lot more than the peak level, but bass-heavy productions or dense arrangements still seem louder on the meter than they actually are. On top of this, RMS levels can vary wildly: if a song has a huge outburst at the end, it will have a very high maximum RMS level, even though most of the song is much softer. Therefore, RMS normalization alone is not a very good strategy.

BS.1770

The ITU (International Telecommunication Union) has developed a standard for measuring loudness that takes all of these considerations into account. The frequency response of the measurement is filtered (this is called 'K-weighting') to resemble the response of the human auditory system. As a result, the meter is less sensitive to low frequencies. Next, the average RMS level is analyzed, over the course of the entire production, so one short burst of energy won't cause the entire production to be automatically turned down; dynamics are allowed. However, this doesn't mean that you can cheat the system. It's not possible to reduce the average level by using a lot of silence, just so you can make the other sounds extra loud. The meter uses a gate to make sure that silence and

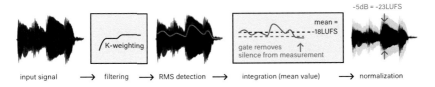

Figure 14.3 The various components used for measuring and normalizing the loudness of a sound clip.

background noise aren't included in the measurement. In other words, this new meter (which follows the ITU BS.1770 standard) measures the average loudness level of the 'foreground sound' of an entire production, and expresses it in loudness units below digital full scale (LUFS). This loudness value can then be used to normalize a production at the desired reference level. For television, this level is −23 LUFS (based on the EBU R128 standard that is now in use in most parts of the world) or −24 LKFS in the US (based on the ATSC A/85 standard, which is basically the same thing). The algorithm used by iTunes (Sound Check) normalizes at approximately −16 LUFS. Not all applications use a method for normalization that's as reliable as BS.1770, but usually the differences aren't more than one or two decibels.

TV

Television is the medium that has made consumers most aware of the existence of loudness differences between programs—especially between the programming and the commercials. If ordinary citizens start to complain about the sound quality rather than the content of the programs, then something must be seriously wrong. And for a long time, that was the case: the only delivery specification was a maximum peak level. Nothing could prevent you from submitting a commercial with a 3 dB dynamic range and enough 3–8 kHz in the balance to wake up every sleeping dog on the block. This advertisement would then be blasted into the living room at two to three times the loudness of the previous program. Fortunately, the delivery standard is now adjusted to −23 LUFS, according to the EBU R128 specification. This allows for a pretty large dynamic range, and in many cases there's no need for a final limiter when you're mixing or mastering for television.

Aim Well

If you know which medium you're mixing for, you can take the reference level of the normalization into account. This way, you avoid having to use more compression and limiting than necessary in order to reach the reference level without peaks exceeding −1 dBFS (True Peak). This generally has a positive effect on the definition, spaciousness, and even the loudness of your mix. Indeed, with the new system, more peaks mean more loudness instead of less. Of course, peaks are only partially responsible for the perception of loudness—the average level is much more important—but if you normalize at a certain LUFS level, the mixes that use the full space available for the peaks are generally louder than the mixes that are limited at a lower peak level (see Figure 14.4). For most music, the −16 LUFS used by iTunes (and recommended by the EBU) or the −14 LUFS that Spotify and Tidal use are very reasonable guidelines, unless you're certain that even smaller dynamics serve a musical purpose. If that's the case, you can put as much limiting on your mix as you want, of course. But if you hear it again in a playlist, don't be surprised when it sounds a bit softer in absolute terms than more dynamic material.

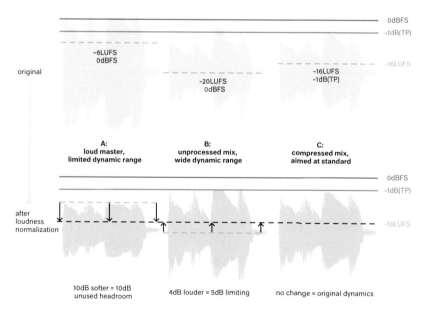

Figure 14.4 Different masters respond differently to loudness normalization. The top image shows the old situation, with large differences in loudness between the three fragments. After normalization, all the fragments are equally loud, but C sounds most natural. Because the sound of fragment A is more restricted than the rest, it seems a bit duller and less loud. Fragment B had to be limited to meet the standard, so it doesn't sound as good as C, which didn't require any manipulation.

More than a Number

No matter what standard you use, two mixes normalized at the same LUFS level won't necessarily be equally loud. Since there are still variables that can make one mix feel louder and more powerful than the next, it's worth your while to learn how to control them. The nice thing about these variables is that they always do their job. No matter where and how loud you play the mix, if its musical energy is well positioned, this energy will always be reproduced. In this case, 'well positioned' has more to do with the frequency balance and arrangement than with dynamics. An empty arrangement with a lot of energy in the high midrange and little reverb can sound extremely loud and close. Furthermore, this arrangement has an advantage in terms of LUFS measurement, because it only needs to divide the available energy among a few elements. For an eight-piece band recorded at a distance, the measurement is not so favorable, but if the positioning of the energy is good, it will still feel loud enough. However, this does point to the shortcoming of any normalization system: it can't distinguish between musical contexts. As a result, small acoustic

---

Radio

Radio also uses a form of loudness normalization, albeit of a less elegant kind than the television version. Through the use of blunt force, a processor (usually an Orban Optimod) brings the dynamics down to a very small range, for the purpose of attracting the attention of channel surfers, and minimizing the information that's lost, for example due to the background noise of a car. For music, this method has far-reaching consequences: besides the fact that all those empty intros are now louder than the choruses, some productions have a much more powerful sound than others. In that sense, radio processing actually resembles the ITU BS.1770 normalization: a greater dynamic range is rewarded with more perceptual loudness. Productions with very compact mastering that still sounded acceptable on CD can completely lose out to dynamic productions on the radio. I once attended a test conducted by Ronald Prent and Darcy Proper (in collaboration with Orban) where this became painfully clear. After five different masters of the same mix—with increasing absolute loudness levels—were sent through the Optimod processor, the original mix came out the best. This is really an upside-down world, because without processing, the first two masters definitely sounded better than the mix.

songs always sound too loud compared to grand arrangements, but it's still a big step forward from the old situation. Today, the 'musical loudness differences' are usually less than a few decibels, while before this could easily be twelve decibels or more.

References

The mother of all questions when it comes to loudness is of course: is my production missing the mark? This fear is what once triggered the loudness war, but despite the new normalization principles, it's still just as relevant. In order to not miss the mark, today it's mainly important to create a mix that has enough definition and that doesn't waste any energy on frequencies that don't matter. You could say that these are the main qualities of any good mix, so that's convenient. But how do you know if you're at least getting near the mark in terms of impact? To answer that question, you'll have to use the same LUFS level to compare your own production with reference tracks. This can be done by normalizing your own production and a number of reference tracks with a batch normalizer. But also during mixing and mastering, you can already use a LUFS meter and aim for the normalization level of your reference tracks. And nowadays you can even find tools that compensate for loudness differences in real time, so you don't have to do the normalizing yourself. This is a bit less accurate though, as these tools don't take the entire length of your production and reference tracks into account. Instead, the measurement and compensation are based on a short sample.

But what if you send your mix to a client or label, and they compare it in a player without normalization? Then you have a real problem, because the new standard might allow for an optimal sound, but inherently it also makes your mixes and masters less loud. As a result, all the reference material that came out between 2000 and 2015 (unfortunately including a lot of remastered older productions) plus a part of what's still coming out today will be louder than what you're submitting. The solution is to advise your clients to import your work in iTunes (with Sound Check turned on in the preferences) or Spotify, and compare and judge it from there. This is not a bad idea in general, as most consumers will hear your music with a similar kind of normalization on it—whether it's through a streaming service, on the radio or television, or as a paid download. At most, as an extra service—as long as loudness normalization is not the standard anywhere—you can create an additional version with more limiting.

Problem Solved?

Loudness normalization is not the holy grail that will solve all your problems. The most important thing is still to deliver solid mixes with

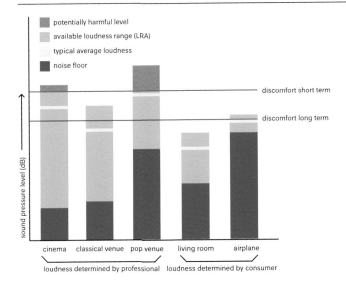

Figure 14.5 Different purposes require different dynamic ranges. If you know the available LRA you can target your mix at a particular situation (for instance if you know it will only be played in cinemas, or if you purposely create the mix live). But if you don't know whether your mix will be played in an airplane or a living room you have to make an educated guess. It's no use anticipating worst-case scenarios by restricting the dynamic range of your mix, as this will ruin the listening experience in all other situations. It's better to aim for a good average experience, which is also dependent on the genre and audience type. Hopefully in the future the systems designed to operate in difficult circumstances (cars, airplanes) will incorporate a dynamic range reduction system of their own, next to loudness normalization.

dynamics that fit the music, audience and listening conditions. The fact that you could theoretically use a huge dynamic range doesn't mean you actually should, as you won't be doing television viewers or earbud-wearing commuters a favor. To put you on the right track when it comes to dynamics, many loudness meters also display the loudness range (LRA): the variation in loudness of a sound clip. In a cinema, you can get away with a larger LRA than on a mobile phone (see Figure 14.5). But whatever dynamic range you choose, sacrificing sound quality to gain loudness is no longer necessary today, and it even works counterproductively. In this respect, I'm looking forward the future, because now it's about other ways of making an impact again, instead of purely about who's the loudest.

Chapter 15

Mindset
A Competition against Yourself

Over time, the technical execution of a mix becomes dead simple. I discovered this when I lost three mixing projects due to a broken hard drive. I had saved the stereo mixes on a different drive, but because I still needed to make some adjustments, there was no other option than to mix the three songs again. Disheartened, I started over, thinking that I'd need another day and a half to get everything back to the same point as before. But to my surprise, after a Saturday morning of mixing, the three songs were done. It felt like I already knew the solution to the puzzle and the way to get there, which allowed me to perform all the necessary actions on autopilot. When I compared my later mixes to my earlier work, I could only conclude that I had done even better on the new versions. Ever since, I have never been afraid to close all the faders and start over when a mix doesn't work. The time is never lost. Because you've learned a lot about the music in the process, you can get back to the same mixing stage in no time.

Up till now, this book has mostly discussed the technical execution of mixing. But technique alone is not enough for a good mix. It can only help you to execute your ideas with more speed and focus. The essence of mixing is about how you form and develop ideas. It's about how you keep track of them, or sometimes let them go. This chapter deals with that side of mixing. It's about the mindset you need to get through the endless puzzle and still be excited to get back to work the next day.

15.1 Doubt Is Fatal for a Mix

Nobody who works on a music production can be certain about everything, as it's impossible to predict the outcome of all the variables beforehand. This unpredictability creates room for chance and experimentation, and that's exactly what makes working on music so much fun. The way you make use of the room you're given—and the uncertainty that goes with it—is inextricably linked to mixing. Uncertainty goes hand in hand with doubt, and that's where the danger lies. However, this doesn't mean that doubt has no part to play in the mixing process. On the contrary: calling your decisions into question is essential if you want to make good choices.

While exploring the material, brainstorming with the artist and producer, shaping your own vision and quickly trying it out, there is room for doubt. During this exploratory phase, doubt is the engine that drives experimentation. It keeps you sharp and stops you from doing the same thing over and over again. You're looking for an approach that feels good from the get-go, which also means turning down the options that don't immediately feel good. This might sound shortsighted, but it's necessary, as there are simply way too many possibilities to thoroughly consider. It's better to trust your intuition and go for something that feels good right away.

Once you've decided to go with a particular approach, the exploratory phase is over and—at least temporarily—there is no way back. From this moment on, doubt is dangerous. You are now entering the execution phase of the mix, which requires complete focus. You work quickly, decisively, and without looking back. It's okay to still have doubts during this phase (after all, you're only human), but only about the execution of the plan. These doubts are very welcome as well, as they force you to be precise and fully optimize the plan. But as soon as the doubts start to spread to the plan itself, you're lost. Mixing is pointless if you don't know what you're going for. If you change the original plan, you throw away almost all the work you've done up till that point. To avoid this, you'll need to let go of your doubts while working on your mix, and take off the training wheels of external input, reference tracks and rough mixes. You'll have to fully trust your own ability and the material you've been given. But what if it turns out that your plan doesn't work? Well, I guess you'll have to cross that bridge when you get there. It's pointless

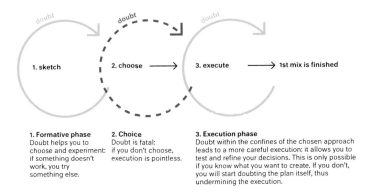

1. Formative phase
Doubt helps you to choose and experiment: if something doesn't work, you try something else.

2. Choice
Doubt is fatal: if you don't choose, execution is pointless.

3. Execution phase
Doubt within the confines of the chosen approach leads to a more careful execution: it allows you to test and refine your decisions. This is only possible if you know what you want to create. If you don't, you will start doubting the plan itself, thus undermining the execution.

Figure 15.1 Doubt is often useful, except if you have to decide on the direction of the mix.

to reflect on half-finished products—do something with conviction and see it through or don't do it at all.

It sounds quite rigid to approach a creative process like mixing in this way. But if you want to get anything done at all, and get around to the choices that will really make a difference in a piece of music, you can't get stuck too long on choosing a direction. Each time you have to go back to the drawing board to adjust the original concept is actually one too many. If necessary, just take the rest of the day off and try again later. Or start working on another song instead. I would even go as far as to say that making a less ideal choice is better than looking for the perfect one for too long. After all, you have no idea what the implications of your choice will be at the end of the ride.

Sometimes you just have to start and don't look back to see if you had it right. If you beat around the bush for too long, you're wasting all your enthusiasm and objectivity on a concept that may or may not work. So it's always a gamble, no matter how much time you put in. This is exactly what separates experienced mixers from beginners: they don't necessarily make better mixes, but during the formative phase they're quicker and more focused in making choices based on gut feeling.

Decisive Action

Mixing has nothing to do with unlimited creativity. It's creativity within a context, and you're expected to deliver instant results. In a way, it's a

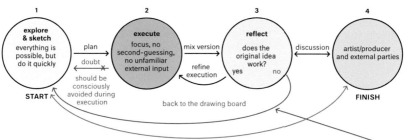

Figure 15.2 There are a few moments for reflection and discussion with your client, but during the execution of the plan there's no room for distraction or looking back. For live mixers, it's self-evident that there can only be discussion during the sound check and after the concert, as it requires their full focus to successfully complete the execution. Ideally, it should work the same in the studio, but don't expect all parties involved to also realize this. If you don't pay attention, the execution phase will be used for feedback as well, which is disastrous for your focus. Therefore, after the preparatory phase is over, you'll have to consciously create room for the execution.

Live

If you have a hard time making pragmatic choices, it can be very inspiring to mix live concerts. The feeling that you only get one chance to make it work will dispel your doubts in no time—there's simply no time for doubt—and help you to let go of the idea of perfection. A mistake is not a disaster, as long as the big picture is still convincing. However, you shouldn't think that a studio mix is the same as a live mix, as a studio production requires a more precise execution. But during the early stages of the mixing process, perfection is a ball and chain—and meaningless at that. If you first make a sketch like a live mixer, the time you spend perfecting the sound won't be wasted. You need to put all the important things in place first, otherwise you won't know what you want to perfect.

lot like playing improvised music. The context is clear, the agreements have been made, you know your instrument, and the rest will become clear within a few hours. If you want to mix on a regular basis, your mixing process should be reliable first and foremost. You have to be sure that there will be something useful at the end of the session, even if it's not your most brilliant performance. Therefore, experienced mixers are good at making decisions: at a certain point in the process they choose a direction and follow it through to the end. There are two challenges when it comes to this: first of all, your choice shouldn't be based on a wild guess. The more you know about the music, production and artist, the more certain you can be of your choice. And second, you want to choose as quickly as possible, because this means more time for the execution of the mix, plus your focus and objectivity won't be so low that your work becomes unreliable. You've probably noticed that these two challenges contradict each other. You want your choice to be as informed as possible, but if you postpone it too long, you won't complete the mix successfully before your ears get tired.

The solution is to learn to sketch as quickly as possible. After all, the only way to know which ideas have potential and which ones can be written off is by trying them out. The setup of your studio and mix template has a huge influence on your working pace, and so has your repertoire of 'universal solutions' for common mixing problems. Your sketches won't just help you to make decisions, but also serve as a means of communication. An audio example will tell your client more than a thousand words. Through efficient sketching, you can get to a mixing

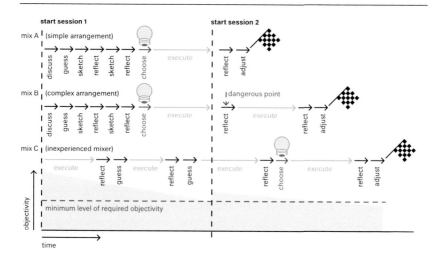

Figure 15.3 The time during which you can stay objective, focused and creative is limited. That's why it's so important to structure the first part of the mixing process well. Executing and refining the mix can still be done after you've lost your focus a bit, but devising and testing new concepts is no longer possible at the end of a session. Complex mixes are a challenge, because when you call it a night and continue working the next day, there's a greater risk of questioning your original concept. The earlier and clearer you define your vision for the mix, the more reliable the process will be.

vision faster, while reducing the risk that it falls short, or that it clashes with the producer's vision (see Figure 15.3).

Be Nice to Yourself

No mix is perfect, and no mix is ever 'done.' You could keep working on a mix forever, but at some point you have to decide that it's finished, if only due to a deadline or a limited budget. I wish it were different, but that's the reality of mixing. Live, this is much easier to accept, as you simply can't control everything. But in the studio, it can be frustrating: you'll always hear aspects in the music that could still be improved.

Your perception can be a burden, because in reality there is a point when the mix is really done. It's just that this point isn't defined by the moment when everything is perfect, because that doesn't exist. The mix is done when the essence of the music comes across to the listener, when all parts in the arrangement fulfill their role, and when the objectivity of the mix engineer has reached its end. If you can't tell anymore if something helps the music or not, you really need to stop.

The latter isn't always easy. Some productions contain contradictions, which means you have to make huge compromises in the mix to still make it work. For me, it's always hard to hear these mixes again later. Even though the mix might have turned out okay, and though it works in the sense that the music comes across, I only hear the compromises I had to make. I'll keep wondering if there couldn't have been a better solution. If I have to make objective decisions about such a mix again (for instance, because the producer still wants to change things), I make sure to protect myself. From that moment on, I use my main monitors as little as possible, and make as many adjustments as I can on small speakers. This way, I'll mainly hear the essence of the music, and not so much the details. It makes it easier for me to preserve the essence of the mix without questioning all my previous decisions. The worst thing you can do during this minor adjustment phase is to smooth out your original approach too much. After all, this is the very approach that convinced yourself, the artist and the producer that it's time for minor adjustments.

As you can tell, we're reaching the point where mixing can be a fight against yourself. The final stage of a mix is the perfect opportunity to start doubting your own abilities, to be intimidated by the way (already mastered) reference tracks compare with your mix, to wonder if your speakers are playing tricks on you, and so on. Add to this the assignments that are called off because clients aren't satisfied with your approach (this happens to the biggest names in the business—mixing is still a matter of taste), and you have a recipe for getting seriously depressed about what you're doing. It's not a bad thing at all to be critical of your work—I don't think I've ever delivered a mix I was one hundred percent satisfied with—but there's a point where it ceases to be constructive. Fear leads to gutless mixes. If you don't dare to push on with your vision for the mix, the result will be reduced to a boring avoidance of anything that could potentially disturb anyone. In the end, not too much of this and not too much of that means very little of anything.

It helps to think like a musician on a stage. Above all, mixing should be fun: you should be able to lose yourself in it and make decisions without fear of falling flat on your face or being judged. If you don't feel this freedom, your decisions will never be convincing. And what if you're all out of ideas for the moment? Then don't be too hard on yourself, and simply trust the people around you. Every day, my clients help me to be a better mixer, by pointing me to things I had overlooked, or by proposing alternative approaches. It's nonsense that you can instantly understand everything about a song. If you want to do your job well and with conviction, you need to give yourself some space. Sometimes you just have to try your luck to hear how an idea turns out, and laugh about it if it fails. Sometimes you have to be brave enough to hold on to your first impulse without getting thrown off by the drawbacks that are bound to come with it. Sometimes you have to admit that the artist and producer are right when they say your mix isn't working yet, and other times you have to trust them when they say your mix is awesome, even though you're not so sure about it yourself.

One thing can never be called into question during mixing: your process. You will need an almost religious belief in it, because whoever said that it's better to turn back halfway than to get lost altogether was definitely not a mix engineer. It's only after the mix is finished that you can afford to think about methodology, and preferably not more than a few times a year. Being biased about the effectiveness of your own way of doing things—and the gear you use—will help you to reach the finish line with a sense of conviction. If you start wondering halfway if the compressor you're using is really the best compressor ever, you could easily end up with a remix. This is why only your mixing decisions can be the subject of debate: whether you use the compressor, and how much. Don't question the fact that you always use the same compressor for a certain purpose. 'You can't change the rules of the game while you're still playing' would be a good slogan for any mixer to frame and hang on the wall.

15.2 Imposed Limitations: Working in a Context

It's tempting to try and emulate the sound of your favorite productions with your mix. But while it's always good to be inspired, it really shouldn't go any further than that. Before you know it, your ideal is defined by the technical execution of another production. When you close your eyes while mixing, you think about your great example and try to approximate it. But what a hopeless undertaking that would be! If there's anything that doesn't work, it's trying to apply the production and mix of one song to another. The guiding principle should always be music that comes across well, and if it sounds good on top of that, you've done even better. The music dictates the technique.

So music that comes across well is not the same as music that sounds good? For average listeners, it is the same: when they enjoy a piece of music, they look at you, blissfully nodding, and say: 'Sounds good, huh?' Meanwhile you, on the other hand, are still annoyed by excessive distortion and dominant bass. This is an extremely important thing to realize: the only people in the world to whom music that comes across well doesn't mean the same thing as music that sounds good are sound engineers. To a lesser extent, this also goes for producers, labels and some artists. You could say that mixers carry the weight of an insider's knowledge about how sound works, and how you can solve problems to make the sound more defined, balanced and spacious. But what if the music needs something else? The question should always come from the music, and shouldn't be prompted by technology.

The best way to prevent yourself from making everything as polished as possible is to deny yourself that possibility. Your best clients—those who push you to your greatest creative achievements—will do this for you. For example, because they have already recorded all the vocals through the PA of a local house of prayer, the production might sound a bit distant and messy, but it's a huge statement. Now you can put all your effort in giving the sound a bit more focus, but it will never be polished. Things would be very different if these clients were to give you all the raw vocals as well. After all, these tracks do sound close and full and warm and nice and technically oh so good. But as you start blending them with the manipulated recording, you stop noticing how meaningless the raw vocals are in this musical context. And that's what it's all about.

Mix engineers deliver their best work when they're severely restricted in terms of artistic direction, because otherwise they will use all the space they get (see Figure 15.4). This means that if the producer doesn't make any decisions to set these boundaries, the mixer will have to. However, this turns you into a kind of eleventh-hour producer, plus you won't get around to the finesse of your mix. Your entire role as a translator of an idea has been shifted to being the co-author of the idea. So now you have to run those damned vocals through the PA of the house of prayer yourself.

It takes time and objectivity to map out (part of) the artistic direction. Afterwards, your ear for what the vocals need in the mix won't be sharp enough for you to still be independent as a mixer. Your qualities as a mix engineer will go down, simply because you've bitten off more than you can chew, and it can be a thankless job to do more than what's in your 'job description' as a mixer. It's for this reason that I would negotiate such an expansion of my employment separately, especially since it could mean running into things that can only be solved by going back to the recording process. On top of this, you'll need to take more time if you still want to be able to approach your mix as an individual project, because a producer is not a mixer, even if he or she is the mixer. If you perform both tasks yourself, it's advisable to keep them separated: first delineate the concept (production) and then translate it (mix).

This scenario happens a lot in electronic music, where the producer is often the composer and mixer as well. This seems like a very simple situation in which you don't have to answer to anyone, but at the same time I can't think of a more difficult way to convincingly get a creative concept off the ground. If no external restrictions are imposed and no frameworks are given, it's very hard to stay on course. With nothing to hold on to, you can keep trying different directions, which can make the production process inefficient and frustrating.

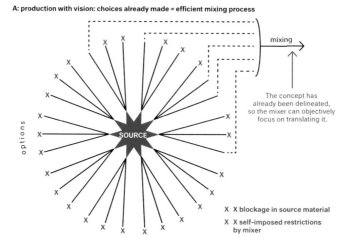

A: production with vision: choices already made = efficient mixing process

mixing

The concept has already been delineated, so the mixer can objectively focus on translating it.

X X blockage in source material

X X self-imposed restrictions by mixer

options

SOURCE

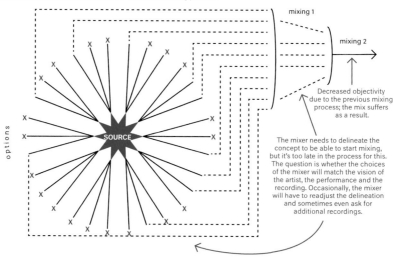

B: gutless production: deferred choices = unreliable mixing process

mixing 1

mixing 2

Decreased objectivity due to the previous mixing process; the mix suffers as a result.

The mixer needs to delineate the concept to be able to start mixing, but it's too late in the process for this. The question is whether the choices of the mixer will match the vision of the artist, the performance and the recording. Occasionally, the mixer will have to readjust the delineation and sometimes even ask for additional recordings.

options

SOURCE

Figure 15.4 Contrary to what you might expect, mixers don't like having a lot of options at all. The lines that spread out from the source material represent the directions you could take with the material. If enough of these directions have been blocked during the production stage, before the start of the mix (the red Xs), your mission as a mixer is clear (A). Deferred choices that are meant to give you more freedom actually frustrate the mixing process (B), as the latter should be about translating a concept, not about creating a concept for music that has already been recorded.

If no one imposes restrictions on you, you'll have to do it yourself. Whether you have to blend eighty middle-of-the-road tracks into a whole that will still sound fresh and different, or build a dance track all by yourself with your computer, the sooner you choose a direction, the better. Choosing also means blocking the possibility of changing your mind. You found a cool sound? Just record it and never look back. You're stuck on a transition that fails to impress? Then remove some parts from the previous section to make it work. You found a sound that can determine the character of the song? Turn it all the way up, so the other instruments will simply have to make room for it.

Do hard and irreversible choices always lead to success? No, sometimes you run into restrictions that turn out different in the mix than previously expected. Some problems won't emerge until you really get into the details of the mix. Then it's time for consultation, and hopefully the problematic parts can be re-recorded or musically adjusted to make the mix work. This method is still very targeted, and it's better if the direction of the production is completely clear and some things still have to be changed later, than having to sort through a myriad of deferred choices, which initially seems very flexible.

15.3 A Guide for Your Mix

It's already pretty hard to choose a direction for your mix, but holding on to it is at least as tough a challenge. A mix is an awfully fluid thing: you change two things, and your entire basis seems to have shifted. That's dangerous, because before you know it, you've forgotten what it was that made the music so special when you first heard it. But changing nothing isn't an option either, as your mix definitely won't get anywhere that way. So you need a way to keep your balance, in spite of the transformations your mix is going through.

Initially, the direction you choose is purely conceptual, and you won't make this concept concrete until the execution phase. In order to do this well—executing a mix without losing sight of your concept—you need a way to capture a conceptual plan. Insofar as the plan contains concrete actions or references, this won't be a problem, but usually the main part of the plan is a feeling or a set of associations that you have formed—mostly as a result of listening to the source material and talking to the artist and producer. This first impression of the music is hard to describe, but it's of vital importance to the success of your mix. It's no surprise that a mix with a concept that's based on how it felt when you first heard the music will work better than a mix that's based on a meticulous analysis of all the details in all the parts. This poses a big challenge, as you have to come up with a way to preserve the feeling you had during the initial phase of your mix, so you can still base your decisions on it at the end of the day.

Be Open

If you want to use your initial feeling as a tool, first of all you have to be open to it. This might sound like something you'd rather expect from a woolly self-help book, but it's necessary. However, this doesn't mean you have to take a Zen Buddhism course, but you do need an empty head, calmness and attention.

The first time you listen to a piece of music should be a conscious experience, so the music will have the chance to settle into your mind. This is why I never play a project in the background while I'm still busy preparing the mix setup. I prepare what I can without hearing the music (routing, naming, sorting and grouping of the tracks, arranging the mixing board, and so on). When that's done, I grab a fresh cup of coffee and quietly listen to the entire production. I prefer to start with a rough mix that clearly conveys the idea. But if this isn't available, I simply open all the faders, and if things annoy me, I quickly turn them down while I keep listening. This is the closest thing to walking into a venue and hearing a band play without prior knowledge, or hearing a song for the first time at a party or on the car radio. If it's a live mix, you could start by having the artists play together on stage without using the PA. This way, you'll notice how they react to each other and who fulfills which role, before you start manipulating this with your mix.

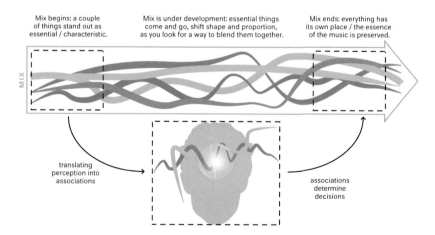

Figure 15.5 Consciously registering your first impressions of a mix in the form of associations makes it easier during the final stage to still distinguish between the main elements (which make the music and mix unique) and the secondary ones (which support the main elements).

Sometimes I take notes, but usually I absorb the music mentally and start free-associating. You can see this as consciously creating connections between your feelings and concrete references. These connections help you to remember your feelings, just like it's easier to remember all the names of a group of new people if you consciously connect them to a physical feature or to someone you know who has the same name. In Figure 15.5, the principle of this way of working is visualized.

Translate

Preserving your feelings is the starting point, but you won't be able to use them as a tool until you can convert these feelings into concrete mixing decisions. In order to visualize how this process works, you could draw a huge diagram in which you would have to display the complete histories of the composer, musicians, producer and mixer. Their fears, loves, influences, origins, upbringing, experiences, character, everything. All of these things converge at a certain point in the diagram and start to affect one another, which eventually leads to all kinds of decisions. Such a diagram would be infinitely complex and almost by definition incomplete. This is why, in the art world, this process is often simply referred to as 'magic,' 'inspiration' or 'creativity.' Great, all these vague concepts for the most crucial part of the creative process. Sadly, it's hard to make these things more concrete, because the translation of your feelings into practice is exactly where the greatest artistic freedom lies. This conversion works differently for everyone, and therefore, by definition, music sounds different when it's mixed by someone else. In that sense, imitating someone else's mix is like trying to fall in love with the same type of person as someone you admire, which is never going to work. Still, there is more to be said about how this magic can come about, as a significant part of it is actually not so magical—it's mainly about estimating what's important for the music, and then drawing on your knowledge of how you could accentuate this in your mix. In these cases, there's a relatively direct connection between your perception, associations and mixing decisions. Another part of it is in fact more abstract, and if you want you can use the word 'magic' for that. Or call it 'chance,' 'revelation from above,' or whatever floats your boat (see Figure 15.6).

Storyline

If all of this is getting a bit too vague for you, or if it's already in your system so you don't have to consciously think about it, there's another concrete and easy-to-implement principle you can use to direct your mixing decisions: all music has a story. It takes you from one point to

If you consciously translate the first feeling you get from a piece of music into associations, it will be easier to preserve this feeling and convert it into a plan that you can hold on to while mixing. This conversion can be direct or abstract. ★

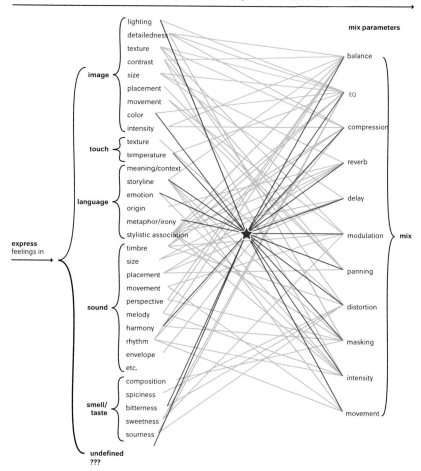

Figure 15.6 The method I use to hold on to the feeling I get from a piece of music when I hear it for the first time. By consciously registering my associations, I create a memory that I can call up later in the process, and then translate it into mixing decisions. Often, a large part of my feeling has little to do with how others perceive the same piece of music. Therefore, it's not my intention to translate it literally so you can recognize it in the music—the conversion is usually much more abstract. I mainly use these concrete associations as a memory aid, to bring me back to the moment when I first heard the music.

another, and things happen on the way. These musical events keep the music interesting to listen to, and therefore it's important to emphasize them in the mix. This is a lot like mixing music and sound for a film, except that the events on the screen are prompted by the scenario, not by the music. Since these cinematic events are much more concrete than when you work with music alone, it's easier to make distinct mixing decisions. When I mix music for movies or live performances, I always find myself having fewer questions than when I only hear the music. The images dictate when what is important; they determine the course of the mix.

If there is no (visual) story, you can also create one yourself. You can come up with a music video to go with the song you're mixing, or play a movie in the background with a mood that matches the music. In an interview with mix engineer Tom Elmhirst (by sonicscoop.com), I read that he regularly plays movies in the studio while mixing, to keep having images to connect with the music. For me, it often works well to create scenarios for transitions that aren't powerful enough yet. This is also a very effective way to communicate with artists and producers about where a mix should go. A songwriter once made it clear to me what had to happen in the transition to the bridge by telling this story: 'You're at a party among friends, but you start to feel more and more alienated from them. Eventually, you take a big jump and dive into the swimming pool to close yourself off from the outside world. Underwater, you become completely self-absorbed.'

Images Are Just like Sounds

Many visual features have a sonic equivalent, and connecting these two can help you to get a better grip on the meaning that you give to a sound. For example, it's very tempting to make everything big and impressive, until you 'see' that a certain element should fulfill a much smaller role in the image, and be anything but prominent. Personally, I find it a lot easier to render the various roles and spatial parameters in a production visible and tangible than audible. Equalizing something very brightly is the same as overexposing a photo, and using compression is the same as adding a lot of extra detail. Many principles from the visual arts about composition apply directly to mixing, and vice versa. The visual connection can also be much more abstract, and take place at an emotional level rather than through very concrete associations.

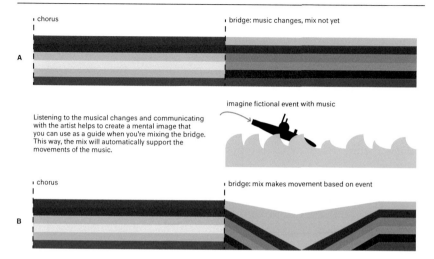

Figure 15.7 Mixing a song based on a (made-up) story motivates you to create powerful, unambiguous movements. A single image of a submarine going underwater can lead to dozens of mixing decisions, in a very intuitive way.

A concrete storyline like this can be a stimulus for hundreds of mixing decisions, all of which serve the same purpose. The result of such a connection is a mix that makes a powerful statement: your goal is to create an event that's told by all the instruments, not a random balance in which the instruments just happen to sound good together (see Figure 15.7).

Your audience usually won't have any idea of the underlying ideas that brought you to a particular mix. Which is good, because if you communicate these ideas too literally, your listeners can't develop their own feelings with the music. Therefore, the goal is not to convey your own associations, but to use them to give your mix a certain structure. The audience can feel the difference between a structured mix and one that results from arbitrariness. A structure can really give a mix a distinct character and a narrative quality, which is more likely to evoke feelings in the listener than something that only sounds nice but that's pretty randomly put together. You can compare this principle to an abstract painting: you can give the viewer too much, but also too little structure to hold on to.

15.4 Ways to Stay Fresh

Take Breaks

The longer you can stay objective about your own work, the greater the amount of reliable work you can get done in a single day. In this case, 'reliable' means that you and your clients will still be satisfied with it the following day. Besides training your ability to concentrate (by mixing a lot), taking breaks is the best way to expand this window of objectivity. But it's not as easy as spending twenty minutes at the coffee machine every two hours, like they do at the office. An ill-timed break can also make you lose your grip on a project. Simply taking a hike in the middle of a process or a train of thought is not an option, as this sometimes means you have to do the whole thing again after you come back from your break. A break can help you to take some distance from already completed subprocesses, so you can evaluate them with fresh ears. For example, your first mix balance won't mean anything until it's done. It's pointless to reflect on it halfway, as this will only allow the doubt that rears its ugly head after each break to sneak into your process. It's for this reason that I always have lunch at different times, depending on where I am in the process. I should add that to me, a break means no music, so no reference tracks by other artists either. I don't let the artist have another listen to the mix while I'm taking a break, and I don't perform minor edits that are still on the wish list. Preferably, I spend some time outside the studio, in different acoustics with different ambient noise. I would almost consider picking up smoking . . .

Distraction

One of the reasons to take breaks is that your ears get used to the sound they have to cope with (see Chapter 2). This is what makes a piercing guitar pop out of the mix the first couple of times you hear it, while later it seems to blend in just fine, even though you haven't changed a thing. This kind of habituation creeps in when you listen to the same sound for an extended period of time. So even if you don't take breaks, it's a good idea to not constantly play music. All the work I can do without sound coming from the speakers, I do in this way. For naming, cleaning up tracks and routing you don't need any sound. For editing you do, but you don't always have to hear the entire mix. So this is one of those rare cases where the solo button does come in handy: when you're removing noises or clicks from an individual part, it's fine (or even better) to do this by isolating the track.

Switching to a different type of speakers can also help you get a fresh perspective on your mix, although it can also be confusing if you don't know exactly what to expect from a particular speaker. A less drastic solution that will still work very well is briefly switching the left and right channels. This way, many instruments are temporarily located somewhere else, so it will be more like hearing them for the first time. If the piercing guitar from the example was originally panned slightly to the right, its shrillness will stand out more when it suddenly comes from the left.

It's important to be alert when you listen to the result, because after you've heard your mix twice with an inverted stereo image, the novelty has already worn off.

You can also try to reset your ears with other sounds. And I don't mean intimidatingly good-sounding reference tracks that have everything you'd want your own mix to have, but a completely different sound. This won't keep your ears from getting tired, but it can rebalance your reference of the frequency spectrum. After you've watched some random nonsense on YouTube, maybe you'll suddenly hear how piercing that guitar really is.

Let Your Mind Wander

As much as I value silence when I'm not consciously working on the mix, sometimes listening with half an ear is the best way to signal problems. If you let your attention wander to your inbox, a television show or another source of distraction, you can suddenly notice a huge problem that was right under your nose all this time. All-important realizations such as: 'If I turn the vocals down, the entire mix will fall into place,' will sometimes only come to you after you create some mental distance from the mix. There's a fine line between half-attentive listening and not listening at all, so it's important to find a way that works for you. Personally, I become too absorbed in email and messaging to keep my thoughts open enough for the music. But checking a website or watching sports with one eye (and without sound) works perfectly for me.

Pragmatism and Deadlines

Creativity isn't bound by time, or at least according to people who obviously aren't mix engineers. It's true that good ideas are difficult to force, but what you can do is leave as little as possible to chance. The day that I don't finish a mix because I'm passively waiting for a good idea is the day that I'm replaced by another mixer. A large part of the mixing process consists of structuring and executing ideas that have already been conceived by others, or that are already part of your own repertoire. You can always fall back on these if you're all out of ideas. Because you work so methodically, with tools and concepts you're extremely familiar with, this creates a pleasant state of peace that allows you to be creative. The almost autistic nature you need to have as a mixer—and the desire to leave nothing to chance and always structure everything the same way—makes you confident enough to explore new creative directions. Meanwhile, your method helps you to suppress the fear that the mix won't be finished or that it won't work.

Pros and Cons of Mixing in the Box

Total recall, restoring all the settings of the equipment used to create a mix, was a time-consuming job in the analog era, something you preferred to avoid. Mix engineers tended to finish something in one go and live with it. Clients knew that adjustments were often expensive, and only asked for one if it was so relevant that it was worth the money. This balance is completely different now, since total recall is a matter of seconds on a digital mixing system. The advantage of this is that you can still work on another song after you've reached the end of your objectivity. I've found that this freedom requires a lot of self-discipline. For me, moving on to another mix when you get stuck and trying again tomorrow is not the way to go. It feels like you're deferring choices and interrupting the research process that's inherent to mixing. I want to struggle my way through it first, and only take some distance when my idea for a mix has been realized. That's when the computer really proves its use. The artist, producer and I can then listen to the mix again at home, and after a week we'll dot the i's with a couple of tweaks. If you get stuck halfway, it's pointless to evaluate the provisional result on your car stereo—you need a sufficiently developed idea.

Part of this method is setting deadlines. Not allowing yourself to try yet another idea, forcing yourself to move on to the next element and recording an effect (so you can't change it anymore) can all be seen as deadlines. Deadlines force you to make choices, which in turn forces you to be pragmatic. An example would be if you can't get the lead vocals and piano to work together. The vocals are more important, so if there's still no convincing solution after twenty minutes of trying, the piano will have to make room by being turned down a bit. You might think that this kind of pragmatism is the enemy of perfectionism, and that it leads to carelessness. In some cases this could be true, but more often pragmatic choices keep the mixing process from stagnating, plus they make sure that there's at least something to reflect on at a certain point. After all, it's pointless to refine half-finished products—something has to be finished first, otherwise you can't hear where you are.

Some insights require more time than available in a mixing process. It's highly possible that, after mixing the first three tracks of an album, I suddenly discover the method that makes the vocals and piano work together. I can then go back to my previous mixes and make them work even better with some minor adjustments. On the other hand, if I had refused to turn down the piano during the first mix, I would have embarked on a frustrating course in which I could have spent hours looking for a liberating epiphany, completely losing sight of my objectivity. As a result, I wouldn't be able to make any meaningful choices about the other instruments anymore, because I had already grown tired of the music. Pragmatism is necessary to protect yourself from such pointless exercises. The more experience you have, the easier it gets to be pragmatic. You can already predict the hopelessness of a particular search beforehand, and then decisively break it off.

By the time you've attentively listened to a song fifty times, small details will become more and more important to you. It can even go so far that you lose sight of the essence of the mix in favor of the details. When this happens to me, I switch to my small speakers and listen to the 'bare-bones' version of the mix, stripped of everything that makes it sound nice. Usually, that's when I hear the essence again. And if I notice that my tinkering is only making this essence weaker, it's time to call it quits and live with the remaining 'imperfections.' The mix is done.

Chapter 16

The Outside World
Sound Engineering Means Working for Others

As a mixer, you have to be careful not to alienate yourself from the world. I'm not talking about the long days you spend in a hermetically sealed room, but about the risk of losing perspective on what you're doing by spending large amounts of attention on the tiniest details. It's easy to think that the world revolves around the quality of your mix. After all, that's what you're being paid for. But if the quality you provide isn't recognized, it will be of no use to you. Your clients and their audience determine whether or not you have done a good job, a judgment that often has less to do with audio-technical perfection (if such a thing even exists) than you might think. Your mix is nothing more than a vehicle for your clients' ideas, and as long as you're on the same page with them, you can still put a lot of your own creativity into a project. But if your assessment of the people you work for is incorrect, prepare yourself for the worst.

That's why this chapter is about dealing with external parties in the mixing process. This interaction can make the difference between an innovative mix that never sees the light of day due to a lack of trust, and a production that everyone looks back on with pride—and that will generate new work for years.

16.1 Speaking the Same Language

Today, many musicians have studio equipment themselves, the functionality of which can often rival with the system you use as a mixer. The difference is more in the acoustics, speakers and the quality of certain manipulations than in a lack of options. The main advantage of this is that it's easier for artists to test and communicate their ideas. They understand what is and what isn't possible production-wise, and what technique could produce what result. Initially, this leads to very concrete requests for you, the person who will bring their vision to fruition. In spite of their own resources, most artists still need an engineer for a good mix. After all, equipment is not the same as expertise.

But once you've translated a clear starting point into a first mix that can be discussed, the artist's (lack of) expertise can also complicate the conversation. While musicians might know exactly how a particular technique works, the implications of its use for the big picture are often much less clear. For example, the artist hears that the mix lacks size and depth. A logical request for the mixer would then be to use more reverb. But what the artist doesn't know in this case is that the problem is actually caused by an excess of reverb in the mix. This stacking of reflections and echoes creates masking, which clouds the depth image.

This difference in expertise means that a mix engineer has to be able to interpret feedback. 'Less compression' could mean both more and less compression, depending on the context. A comment about panning could just as well apply to balance or EQ. Lots of conflicting remarks from different band members about the same passage usually means that the passage doesn't work as a whole yet, and there are probably several ways to solve this. Therefore, you should take comments from clients about your work very seriously, but not too literally. In that sense, it can also be an advantage if an artist steers clear of technology and the jargon that goes with it. The statement 'the chorus isn't really taking off yet' will tell a mixer infinitely more than the well-intended remark that the piano in the chorus could use a bit more compression. It's better if artists keep speaking their own language to convey their message, but some of them seem to think that the mixer won't understand them that way. Therefore, as a mixer, it's extremely important to immediately remove any possible barriers that keep artists from expressing themselves.

It's for this reason that I try not to talk to artists about technology, no matter how much of an audio geek I actually am. Instead, I talk about impact, rhythm, leading and supporting roles, temperature, colors, storylines, tension, texture, contrast, and every metaphor I can come up with to describe the sound. This way, I let the artist know that I'm not only thinking in frequencies and decibels, plus it's a way for me to find out what type of language connects with the artist. Sometimes it turns out that tech lingo actually works well, but more often I find that abstract concepts have a better effect. From that moment on, I will consistently use these concepts—and variations on them—in our communication. This creates a primitive 'common language,' which is essential if you want to bring a production to a good end.

It's very easy to think you're communicating well when the artist mainly agrees with you because your story makes sense—and moreover, because you have more experience. If you don't test the language, you'll never know if it actually means the same to both parties. It will be too easy for you to push your own vision, as the artist doesn't fully understand what you mean. At first, this will feel like you're being given a lot of freedom to add your own signature to the production. But no signature of a producer or mixer can be more important than the artist's signature. And often the critical questions of the artists—and usually those of the audience as well—will still come after the process, which means they won't come back as clients and as listeners. This is exactly what makes it so hard to do mix pitches, 'trial rounds' in which several competitors create a mix of the same song, and the winner gets to mix the entire album. Of course this will tell the artist something about your skills, but due to the anonymous nature of this construction, you'll have to magically predict which direction the artist wants to go with the production. In a good partnership, the artist and production team communicate on the same level, feel free to criticize each other's choices, and ultimately push each other to new heights.

Making interim versions and taking the time to reflect on them is a way to find out if an artist's idea of 'blue' is the same as yours. It sometimes seems unnecessary, if a mix is all smooth sailing and you easily reach a point where everyone gets excited. But don't forget that judging mixes is not an everyday activity for artists, and that they're presented with so many permutations of the same song throughout a long studio day that their objectivity is sometimes all used up by the end of it. Therefore, I always encourage artists to listen to the mix again at home, in their own familiar surroundings and with fresh ears. If they're still excited about the mix when they return to the studio, and notice that you take their observations seriously—or even better: if you've also reflected on your work yourself, and hopefully identified the same points for improvement—you can build a relationship of trust, so you'll only need half a word in the long run.

Interests

A one-on-one relationship with an artist is rare. Usually there are many other stakeholders involved in a production, who all want to have a voice in the mixing process. In a typical scenario, the songwriter, musician(s), producer(s), record label and management all provide commentary on a mix, and not necessarily in agreement. Since the mixing phase is the moment when a production first reveals its final form, it's usually also the ultimate opportunity for stakeholders to find out if their expectations of the production are being met.

In the best-case scenario, all parties agree on the most important choices, but sometimes disagreements about the artistic identity of a production can suddenly come to the surface. If the producer expects you to make a dark-sounding mix with the vocals buried mysteriously under heavy instrumentation (following all the sketch versions up till then), while the management hopes that the mix will make this darkness bloom into a poppy singalong song with the vocals firmly in the front row, you have a problem. This extreme contradiction is too big to be ironed out with a middle-ground mix, as you would end up in a meaningless no-man's-land in between, with a weak compromise. Your mix now serves as an indicator of the conflict that the producer and management need to solve between themselves before you can continue. When this happens, it's very important that you don't make your own role bigger than it is, no matter how attractive it might seem to attribute a great artistic importance to the mix. Before you know it, the problem is blamed on the mix, and you fail to please either side. It's better to indicate where the boundaries of your mix are: you can put the artistic identity of a production in a certain light, but you can't change this identity in the mix (unless you're asked to do a remix, which would make you the (co-)artist yourself). This way, you force the stakeholders to first agree among themselves before they present you with a new challenge.

Fortunately, most contradictions are less extreme in reality, and by assessing all the interests and cleverly responding to these, you can usually please everyone involved. Sometimes, a good talk that demonstrates your concern for the 'marketability' of a piece of music is all you need to get the management excited about your mix. Evaluating a mix is a highly

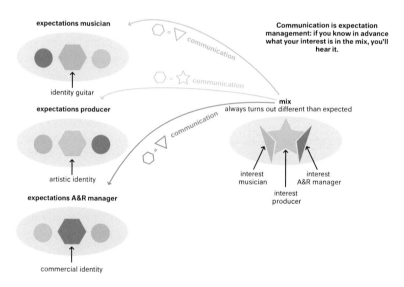

Figure 16.1 All the stakeholders in a production will have prior expectations about their own interests. For them, it's important to understand that the mixer takes their interests into account, even though they might not hear it in the mix—which always sounds different than expected. This is why communication between the mixer and the stakeholders is essential to give the mix a fair chance.

subjective thing, and if you can 'prime' the stakeholders by making them understand your decisions, it can significantly improve their perception. In a way, you make them look at the mix through your eyes, while re-assuring them that you have their interests in mind. A guitarist whose part is considerably turned down half of the time will be disappointed to hear your mix, since many musicians are primarily focused on their own interests. But if you explain during mixing why you don't want to give the guitar the same role throughout the song, by presenting your idea for the big picture, the guitarist will listen to your mix differently (see Figure 16.1). It goes without saying that your mix should also work on its own, for people who weren't influenced by you in advance. The priming I'm talking about only concerns those who already have expectations due to their interest in the production—as opposed to unsuspecting listeners.

The most blunt form of priming is intimidation with a thick CV, a strategy I never use with artists, but which is sometimes necessary to get commercial parties on board. In order to win some people's trust, the smell of success that surrounds your name can weigh heavier than the actual mix you're making.

After all these political games, everyone involved can still have negative opinions about the mix, all of which will have to be heard. I always apply a strict hierarchy when I implement external comments. The artist and producer have the casting vote, and the main challenge for me is to make as few concessions to their vision as possible, as I try to harmonize it with the other requests.

16.2 Reference Tracks

Talking about sound is difficult. Finding the right words to express what you want to say is one thing, but finding words to describe sound that will mean the same to someone else is a challenge. It often proves much more effective to use examples: if you can actually hear what you're talking about, miscommunication is a lot less likely. Still, I usually try to avoid using pieces of music other than the subject of the conversation, because it's extremely difficult (and often even impossible) to isolate the parameters you're talking about from the rest of the music. For example, a bass sound is never a stand-alone entity. Besides being the sum of a range of unique variables in the recording process (composition, perform-ance, instrument, microphone, acoustics), it's also the product of its context: the arrangement, the role division, the production approach and the mix. Therefore, a certain sound that works great in one production might not work at all in another song. The danger of referring to something as specific as a bass sound is that you start to lose sight of the musical function, and focus too much on the characteristics of the sound itself. Most listeners don't care about the sound of the reverb you use, but they do care whether or not it helps to convey an emotion within the specific context of a particular song.

Conceptual Phase

Now I make it sound as if reference tracks have no place in the production process, but it's not as simple as that. Analyzing reference tracks is a great way to learn from others and gain inspiration. When deciding on the direction of the production—the conceptual phase—reference tracks can help to provide the framework for your own production to relate to in a certain way. They're the perfect catalyst for a substantive dialogue (even if it's just you talking to yourself) about what the production should and shouldn't be. This way, you're not copying specific elements from the examples, but you use them to get a clear idea of the feeling you want your own work to give off. For example, the kind of subdued tension that you get from a reference track can be a good starting point. But the exact way in which you will evoke this kind of tension in your own production won't become clear until you start working on it. By referring to concepts instead of literally quoting references, you prevent yourself from focusing too much on the specific sound that should create this tension, a sound that might not fit your own production at all.

Avoid Comparison

There are several reasons why I don't use reference tracks during mixing. First of all, they are too intimidating: it's as if you've just finished building the basic structure of your new house, and you start comparing it to the fully furnished mansion you've always wanted to have. This can have a demotivating effect, and it can make you doubt the foundation of your mix, which is very dangerous at this stage. After the conceptual phase, mixing requires an almost naive faith in your own concept and skills, otherwise you'll never bring the mix to a successful end.

Another problem is that as you get deeper into the mixing process, your focus shifts from the sensation of hearing something for the first time to the details of individual sounds. If you switch to a 'fresh reference' at this point, it will make a completely different impact than your own mix. This is not only because the reference track sounds different, but also simply because it's new. After listening to it for hours on end, your own mix can never be as pleasantly surprising to you anymore, because by now you've heard it eighty times.

Finally, during mixing it's hard to keep hearing the difference between the artistic concept (which reference tracks can help to define) and the mix details that are different for each production. During the conceptual phase you could still look at these from a distance, but now the details

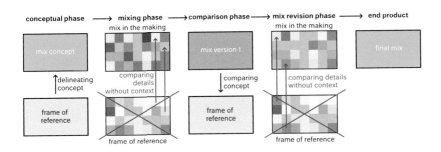

Figure 16.2 While mixing, your focus shifts to individual details, and the comparison of your work with reference tracks leads to confusion. During the moments between the actual mixing, you can compare the impact of the overall mix with your frame of reference.

have suddenly become extremely important. Whether you like it or not, this means that you can't compare the reference tracks with your own production in the same way as before you started mixing. Suddenly you notice how different your bass drum sounds. But is that good or bad? Is the sound of the bass drum in the reference track important for the artistic concept? During mixing, there's only one way to answer these questions, and that's by taking the training wheels off. Does the bass drum work in the context in which you've placed it, and is it in line with the artistic concept you have in mind for this mix? If so, it's pointless to give it any more thought.

The Moment of Truth

As soon as a mix is done—or at least the first version of it—it can be useful to compare it with the outside world again, but this doesn't mean that it's easy. Before you know it, you're back in detail mode, skipping back and forth between one chorus and the other to hear if there's just as much bass in it. This kind of comparison without context is something you want to avoid, as the amount of bass says nothing if you haven't heard how the bass relates to the whole. In this case, 'the whole' means the story of the song from intro to outro, not just the relation of the bass to the other instruments in the chorus. The best way to judge if a song works is to hear it front to back 'in the wild.' By this I mean that listening conditions should be less controlled than in the studio: if you can't hear the details so well, you'll automatically be less focused on them, and you'll hear the big picture instead. That's why the car is such a popular place to judge mixes. If you then also engage in another activity while you're listening—like driving, doing the dishes or having a chat with friends—you're much more likely to notice whether or not your mix succeeds at foregrounding the essence of the music. I write down the observations I make during such a listening session, so I can get back to them later in the studio. I ask the artist and producer to do the same, and from all these impressions I distill a common line for adjusting the mix.

Fair Comparison

During this phase, it's also important to test how the mix relates to the reference material, but make sure to take the influence of mastering into account. I used to always send clients versions with a peak limiter on the mix, so they wouldn't hear a dramatic difference in loudness with the sometimes heavily mastered reference material they used as a touchstone. A slight difference in loudness can be perceived as a huge difference in fullness, depth, clarity and impact. And simply explaining that any loudness differences would be straightened out in mastering wasn't enough to give some artists the confidence that the mix was already powerful enough. Nowadays this process is a bit easier, thanks to the loudness normalization that's built into many digital music services (see Chapter 14). Now you can advise your clients to use a music player with this feature (for instance Spotify, or iTunes with the Sound Check option turned on) to make a fair comparison between the mixes and the reference tracks. This saves you the 'premastering' you put on your mix at the end of the day when your ears are already tired, which sometimes does more damage than you initially might think.

Even at the same loudness, it's still difficult to compare mixes and reference tracks. Comparison theory from the field of psychology can provide insight into this, as it describes how a previous event can influence your current perception. Dip your left hand in cold and your right one in warm water, then put them both in the same lukewarm water and voilà: your left hand will seem warm and your right hand cold. But it goes further than that: if you first browse through a magazine full of attractive men or women, a random group of men or women will seem less attractive to you than if you hadn't picked up the magazine. For-tunately, your image won't be permanently distorted (this effect has a 'release time' as well), but it's this very mechanism that allows fashion trends to perpetuate themselves: everyone looks at each other and tries to outclass the 'competition.'

Music isn't meant to always sound the same, so when you compare different productions, it's important that the intentions relate to each other in a meaningful way. A singer-songwriter with a delicate and personal song should sound more intimate and close than a DJ who fires up the dance floor with endlessly repeated lines. And on this dance floor, the beat should be more dominant than the low end of a blues band.

Metal should be ear-splittingly loud, even at a low playback level. Therefore, it's important to compare your mixes in context: if you first play an entire punk album and then put on your own jazz mix, it's no surprise that it will suddenly sound warm and intimate. But will it still come across this way in a playlist with other warm-sounding music? You can easily check this with a loudness-normalized streaming service like Spotify, by finding a playlist that could potentially contain your own mix, for example 'Country Pop Hits.' Next, you normalize your own mix at the loudness level used by the streaming service, and add it to a local copy of the playlist. It works the best if the list is in shuffle mode, so you'll be surprised when you suddenly hear your own mix.

16.3 Feedback on the Source Material

Ideally, the music seems to mix itself and all the pieces of the puzzle fall into place without any extra tweaking. However, this isn't common practice. Usually, the required tweaks have nothing to do with incompetence earlier in the production process, but with the fact that you can't really know what the combined impact of the separate components will be until you start mixing. Sometimes, problems with timing, intonation, frequency stacking or a lack of power or filling can stay under the radar until the mix brings them to the surface. When this happens, the best mixers will do whatever it takes to still make things work. In that sense, there's no such thing as a well-defined job description that tells you what you can and can't do with a piece of music. Often, the artist doesn't even need to know that it took some tinkering under the hood to reach a convincing mix. You can avoid a lot of anxiety and insecurity if you don't draw attention to the problems you came across. Only if you get requests that you can't carry out due to these problems do you then have no choice but to burst the artist's bubble. For example, you had placed an extra piano part very softly in the background of the mix because it could still add something there—despite its messy timing. The more important you would make it in the mix, the more unstable it would become—but now you get the request to turn it up.

Simply saying no is not a good idea in this case, as this will make it seem as if you're unwilling to change things about your mix. That's why I always take the position of doing whatever it takes to make the artist excited about the mix, but meanwhile I do have my own agenda of making sure that the mix still functions. This might seem like I'm going behind the artist's back, but in reality it's one of the reasons why people hire me. If after a while the artist's focus shifts from only the extra piano part to the impact of the overall mix, chances are that the original request is reconsidered. In case of doubt, it's my job to make sure that the decision that's eventually taken serves the mix as a whole. So instead of denying the artist's request, I demonstrate the influence of the extra piano by making it just a tad too loud in the mix, maybe even louder than what I would have done if the timing had been perfect. This way, the artist won't just hear the piano, but also the influence of the piano on the mix, especially if I point it out. This often leads to a more radical decision (artists tend to be pretty good at making those) than the one I had initially made by turning the piano down: sometimes the artist will scrap the entire part or record it again.

The most important thing about this method is that you try out suggestions, instead of simply saying that they will cause problems. Besides making artists feel that you take their suggestions seriously, it's also a way to test your own judgment. Sometimes it turns out that I was wrong, and a part that seemed problematic at first actually works great when you turn it all the way up. The music isn't mine, so it's an illusion to think that I can correctly interpret the purpose of every single element it contains without any explanation. On top of this, it will be a very different experience for artists and producers to go on a joint quest to discover why something doesn't work, instead of having their 'mistakes' pointed out to them by someone else. This is one of the main reasons why it's good for artists to be present during mixing—especially when you're making adjustments to the mix. Their presence helps to put the responsibilities where they belong: the people involved (including you as the mixer) will get to know the boundaries of the production, and also find out what these boundaries are dictated by.

This way, everyone will eventually understand why the production sounds the way it does. They will see that the mix isn't something that just happened to come about this way (and that can be randomly adjusted afterwards without the risk of losing the essence), but a logical and deliberate result of everything that came before. However, the fact that it's logical doesn't mean that it's easy. Sometimes it takes quite a lot of experimenting and revising before the point is reached where everyone feels that the best way has been found. In longer-term partnerships the dialogue does tend to get easier with time, so you'll feel less inhibited to ask for a musical change directly. But if you bring this up too soon, the artist can get the idea that you don't 'feel' the music, which could result in a loss of confidence. That's why I always try to start from the assumption that each element in a production has a well-thought-out role, unless I'm explicitly told that there are doubts about the arrangement, and suggestions are welcome.

16.4 Feedback on the Mix: Revisions

Even after I had been mixing on a regular basis for three years, it was still a personal achievement for me to accomplish a mix. The result of the long struggle to make sure that all the instruments could actually be heard at the same time was something I had to work so hard for, that I felt the need to defend it tooth and nail. The slightest change would throw the mix out of balance, and it took a lot of effort to get it back to the point where it worked again. In my mind, there was only one mix that could possibly work, and as far as I was concerned, it was ready to be sent to the mastering engineer. But now that it's less of a struggle for me to create a mix that works, I've also come to realize that there are at least a hundred mixes that could possibly work for a particular song. This is why I'm not nearly as persistent as I used to be when it comes to making adjustments to my first version. Often, I even redo entire mixes, for instance if I'm a couple of tracks into an album and I suddenly find the approach the album needs, or if the producer or artist suggests a different angle (for them, it can also take some time before they realize what they actually want to ask of me). But even now, there's a point where I feel that the best mix has been made, and it's better not to change too much. This section of the book is about dealing with feedback after this point—when the mix approach is completely clear, and you have to be careful not to weigh down the foundation of your beautiful house with too many extra furnishings.

Louder Is Softer

The most frequently heard request at the end of the mixing process is to make a particular element louder. Of course that's always possible, but in a good mix there's usually not so much room left to add things without sacrificing the essence. This is because a good mix uses the maximum available space to highlight the music as well as possible. Even if it's a trio consisting of piano, bass and vocals, the balance can be so delicate in the final phase of the mix that one extra decibel for the bass is enough to compromise the brightness of the vocals. But again, it's not a good idea to bluntly deny this request, as you don't want to come across as uncooperative and possessive about other people's music.

What you can do is teach: if you explain and demonstrate that you can't make things louder anymore without making other things softer— but that you're happy to do so, if that's what the client wants—your dilemma will be clear right away. This way, an adjustment won't result in an increase in density that puts too much pressure on the foundation of your mix, but in a shift in proportions that will still fit within the confines of the mix. Turning one instrument up means turning the related instruments down. The purpose of this approach is not to discourage comments, but to make it clear to everyone involved that each request during this phase will affect the basic characteristics of the mix, which they were already satisfied with.

I always try out every request for an adjustment before I judge it, because even at the very last moment it's possible for unexpected gems to emerge. On top of this, I'm not afraid of losing my previous mix anymore, thanks to the total recall option that comes with using a DAW. Some people are critical of the endless possibilities of adjustment this provokes—and the resulting tendency to put off decisions. But at the same time, it has also become much easier to test ideas, without having to worry about losing your original work. If it doesn't get better, you can always go back.

This does require some self-control, as you could endlessly continue making adjustments that mainly cause changes, not necessarily improvements. This is why I have one guiding principle: if there's no problem, I won't change anything. Only if I feel that there's still something to be gained will I start looking, but definitely not randomly. At this stage in the mixing process, you have to realize that you've already explored most of the avenues with the material, and that you're not crazy. The times that I started considering the same choices again, my quest often ended with pretty much the same mix, and it turned out that my original solution wasn't so bad after all. Therefore, I only start this process if there's a truly new idea or insight, which I then work out in detail. This is necessary, because all the elements in a mix affect one another. Trying out a new approach for the drums could mean that you also have to automate the vocals and equalize the guitars again, in order to judge the effect of the new approach.

16.5 Working at a Distance

For artists, the average mixing day is only interesting for a short period of time. You could say that there are only two moments, at the beginning and end of a mix, when communication—first about the approach, later about the execution—is really important. The period in between is when the actual mixing takes place, a long search that an artist shouldn't want to follow moment by moment. Of course there are studios with pool tables, arcade lounges and other forms of entertainment, but you could also just stay home as an artist. It's not that your presence isn't appreciated, but sometimes a long-distance relationship is actually not so bad. A livestream that you can tune into at select moments and from the comfort of your own familiar surroundings is a pretty good way to stay up to date with what's going on in the studio. And more importantly, you'll stay much more fresh and objective than when you can follow the mixer's every detour in detail. With Skype in the background, the communication can be pretty direct—provided that the group that's commenting on the work isn't too big.

The downside of this convenience is the same as why some people get so miserable from using Facebook: everything seems so close, yet so far away. You think that everything is going smoothly and all your comments back and forth are being heard, but it's just a prettier, more polite version of reality that you present to each other. In the studio, you interpret each other's look, body language, the time it takes to formulate an answer and the tone you use to say something. Therefore, in my experience, the decision-making process takes longer at a distance than in the studio. It often comes down to trying and evaluating every single suggestion. In a 'live' situation, it would only take you ten seconds to look each other in the eye and start discussing other ways to crack the nut.

This changes when a project is already in progress, and the common language from the beginning of this chapter has been established as well. Then the disadvantage of distance can even become the advantage of objectivity. If a producer hears something in my mix upon first listening to it through familiar speakers, I tend to trust this judgment blindly. This information is more valuable to me than the same comment shouted from the couch behind me after a long morning of listening to endless mix variations with half an ear, distracted by messages and emails.

In order for long-distance constructions to be successful, it's important to avoid technical issues on the receiving side. For example, the player that the recipients use to open your livestream could have all kinds of sound processing options automatically enabled, like stereo widening and equalization. Besides this, their speakers and acoustics are rarely as good as yours in the mixing studio, but if the recipients often use the same setup to judge mixes in the making, this doesn't have to be a problem. However, if they are only familiar with the sound of mastered material on their audio system, they could be distracted by the character of the mix, which is often a bit rougher. Especially if the system's resolution isn't very high, mixes with a lot of dynamics or low frequencies will translate poorly, while they could sound perfectly acceptable on the same stereo after mastering. But if you were to make 'safer' versions of these mixes, with less low end and reduced dynamics, the final master would lack power and attitude. This is why I blindly accept comments on balance and contrast, but if there are any remarks about the frequency balance or dynamics that I'm personally not so sure about, I ask the long-distance listeners to verify their observations with a good pair of headphones.

16.6 Conclusion

Making music and then entrusting it to someone else is extremely hard. The analogy that creating a work of art can feel like having a baby has become commonplace by now, but this doesn't make it any less true. So no matter how good you are at the technical side of the trade as a sound engineer, if musicians don't trust you to handle their creation with care, you'll be out of work, and they will never know that their baby also feels a little bit like yours.

Chapter 17

Tools
Preconditions for Mixing Reliably

It's not easy to objectively assess the musical influence of your tools. As a mixer, you aren't just a music lover, but also a gearhead. A fantastic new device or a smart piece of software that allows you to perform unique manipulations can make you believe that everything you do with it is beneficial to the music. You could call this inspiration, and personally I don't mind when new equipment makes me try new things. But the ultimate question should always be: does it make the music better? To answer this question as objectively as possible, the tools you use should do exactly what you expect of them. This is particularly true for the acoustics of your (home) studio, as well as your monitors: they should distort the mix as little as possible.

This chapter discusses the tools that you use in every mix: how they work, and why they don't always do what you expect.

17.1 Basic Acoustics

Everything starts with what you hear. If that's a different version of reality, you can't provide objective answers to any questions about your mix, and it seriously impedes the mixing process. Of all the things you can improve to bring the sound you hear as close to the truth as possible (speakers, amplifiers, DA conversion, cabling), acoustics have the greatest influence by far. Mixing on less-than-ideal monitors in a good space works a lot better than mixing on top-notch monitors in a space that sounds out of balance.

17.1.1 Space

A good listening space starts with choosing—or building—a space with the right dimensions, as these will determine (among other things) how the space responds to low frequencies. Three factors are important:

- the volume of the space;
- the ratio of the dimensions (length, width, height) of the space;
- the shape of the space.

These three factors combined influence the pattern of standing waves that occurs in the space. Standing waves have wavelengths that fit an exact-integer number of times within the dimensions of a space, which can cause them to resonate. Through addition and subtraction, these resonances create peaks and dips in the frequency response of the space (see Figure 17.1). Standing waves have a significant influence in the range from 20 to about 200 Hz, so pretty much in the entire bass range of your mix. By optimizing the three factors, you can make the pattern of standing waves as balanced as possible. This means that there won't be clusters of standing waves around the same frequency, but that they will be evenly distributed across the spectrum.

Preferably, the volume of a listening space is at least 90 cubic meters. This way, the effect of standing waves on the music is minimized. It's definitely possible to make smaller spaces workable, but generally,

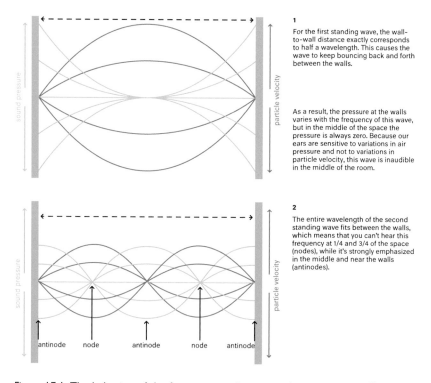

1

For the first standing wave, the wall-to-wall distance exactly corresponds to half a wavelength. This causes the wave to keep bouncing back and forth between the walls.

As a result, the pressure at the walls varies with the frequency of this wave, but in the middle of the space the pressure is always zero. Because our ears are sensitive to variations in air pressure and not to variations in particle velocity, this wave is inaudible in the middle of the room.

2

The entire wavelength of the second standing wave fits between the walls, which means that you can't hear this frequency at 1/4 and 3/4 of the space (nodes), while it's strongly emphasized in the middle and near the walls (antinodes).

Figure 17.1 The behavior of the first two standing waves between two walls.

bigger is better. In a large space, the standing waves with the strongest influence—the ones that fit inside the space exactly once—are lower in the frequency spectrum than in a small space. As a result, the worst coloration takes place below the musical range, with a dense pattern of relatively weaker amplifications and attenuations in the range above. In a small space, the biggest irregularities are well within our hearing range, so they can seriously disrupt the balance of the low instruments.

The ratio of the dimensions also determines the distribution of the standing waves. A good ratio can help you to avoid multiple standing waves with the same resonant frequency, which would make their influence excessive. If you translate this to the shape of a space, you can say that a cube will have the worst response, since the standing waves between the front and back walls, the two side walls and the floor and ceiling all have the same resonant frequency. A good starting point to find out how well your studio passes the test is the EBU recommendation for listening rooms (see box).

Unfortunately, the shape of a space doesn't affect the presence of standing waves—it can only change their distribution a bit. However, the shape can help to move early reflections away from the listening position and avoid flutter echoes between parallel surfaces. (Flutter echo is the sound that keeps bouncing back and forth between two parallel hard surfaces, such as two opposite walls. This creates quick, successive echoes,

---

EBU Recommendation for Listening Room Features

This formula indicates the limits of the ratio between the length, width and height:

$$\frac{(1.1 \times \text{width})}{\text{height}} \leq \frac{\text{length}}{\text{height}} \leq \frac{(4.5 \times \text{width})}{\text{height}} - 4$$

And:

$$\text{length} < 3 \times \text{height}; \ \text{width} < 3 \times \text{height}$$

Finally:

Length, width and height ratios that are integer multiples (within a 5 percent margin) should be avoided. For example: 1:3:4 is not a good idea.

which are clearly noticeable as coloration.) For example, a sloped ceiling would be ideal, because the floor is always horizontal, of course. A little goes a long way: a slant of five degrees is enough to prevent flutter echoes. In some old houses, the walls are already at such an angle! If you plan to build angled walls yourself, try to keep the space as symmetrical as possible, so the stereo image won't be out of balance.

Angled surfaces create converging and diverging zones in a space. In the converging zones, the reflections 'pile up,' as there will be a concentration of the reflected sound between these surfaces. Therefore, a good layout takes into account that these concentrations shouldn't occur between the listening position and the speakers (see Figure 17.2).

A space with favorable dimensions is the most important ingredient of any studio. This means that sometimes it's better to sacrifice some floor space by building a double wall, if this will optimize the proportions of the space. After that, you can perfect it even more with various absorption and diffusion materials.

17.1.2 Optimization

In order to make the acoustics in your studio fit for mixing, you usually try to achieve the following:

- prevent reflections from arriving at the listening position very shortly after the direct sound (from the speakers);
- break up large surfaces to scatter the reflections across a space, making their combined sound less distinctly noticeable;
- damp low resonances to reduce their influence;
- balance the frequency spectrum of the space's reverb.

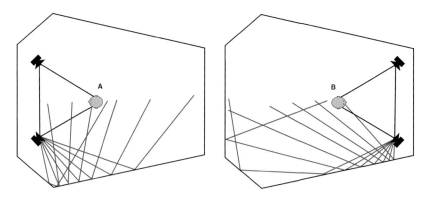

Figure 17.2 Converging surfaces (this can also be a sloped attic ceiling) have a focusing effect on reflections. As a result, the sound at listening position A will be much more cluttered than at listening position B.

Many of the materials needed for this can fulfill multiple tasks at once. Depending on its placement, a thick absorption panel can break up a large surface, remove a direct reflection at the listening position, slightly attenuate low resonances, and also help to balance the overall reverb of the space. In other words: every adjustment affects everything else! To solve this puzzle, it helps to choose a fixed starting point. Therefore, you should first find the best speaker and listening position.

Layout of the Space

You can make a prediction of the optimal position in a space for yourself and your speakers. This prediction is based on the locations of the low-frequency additions and subtractions that are caused by standing waves. But as demonstrated earlier, it's not easy to make these predictions exact. Even professional studio designers still have to experiment to determine the best setup. The trick is to excite the standing waves as little as possible. After all, a standing wave will only resonate if sound energy of the exact same frequency as the standing wave is being produced at the position of the standing wave's pressure differences, which are called antinodes.

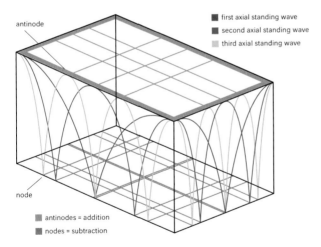

Figure 17.3 Distribution of the nodes and antinodes of the first three axial standing waves in a rectangular space, in two dimensions—the standing waves between the floor and ceiling aren't shown here. Axial means that the standing waves only resonate between two parallel surfaces. Of all standing waves, axial standing waves have the greatest influence on the response of a space—and therefore on the speaker placement. However, in spaces with unfavorable dimensions, the tangential waves (which are reflected by four surfaces on their way) can have a huge influence as well. The locations of the antinodes and nodes in the space are indicated by orange and gray lines, respectively.

A good placement makes sure the speakers don't end up in those places. This is why you should avoid placing them against a wall, in a corner or exactly in the middle of the room—which also goes for the placement of the listening position (see Figure 17.4). Good placement doesn't mean that you won't excite any standing waves at all: the direct sound eventually breaks onto objects in the space and can still end up in the spot where it produces a standing wave. Fortunately, by that time the sound will have lost some of its intensity, and it won't excite standing waves as much as when it comes directly from the speaker.

Armed with this knowledge, you can make an initial setup (Figure 17.5 shows the standard stereo setup) and start by listening to music that you know through and through. Then keep adjusting the position of the speakers slightly, and try to map where the good, and especially where the bad zones are located. Don't forget that the height of the speakers matters as well. You can use masking tape to temporarily mark a place on the floor. While listening, you should mainly pay attention to the low frequencies: once the best spot is found, it's relatively easy to make acoustic adjustments to 'clean up' the mid and high frequencies, but there's not much you can do to change the low end.

In some cases, you can't avoid placing your speakers close to a wall or right above a large table top (or mixing console). This causes speaker boundary interference: long-wavelength sounds reflected off a boundary

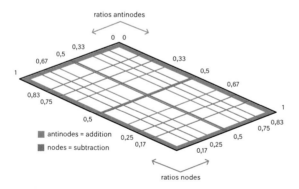

Figure 17.4 A map of the same rectangular space, with the locations of antinodes and nodes indicated by orange and gray lines. It's best to avoid these positions when you're choosing the placement of the speakers and the listener. In each rectangular space, the ratio of the dimensions indicates where the nodes and antinodes caused by the first three axial standing waves will occur. In particular, you should pay attention to the fact that all the antinodes come together at the walls and in the corners. These are clearly places to avoid when setting up the speakers, just like the center lines, where many nodes and antinodes coincide.

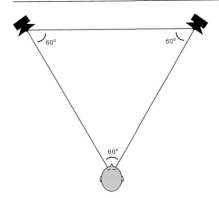

Figure 17.5 The standard stereo setup is an equilateral triangle.

surface arrive at the listener almost simultaneously—and therefore in phase—with the direct sound. A single boundary surface means a theoretical increase of 6 dB in low-range sound pressure. In the worst case, a speaker is on the floor in a corner, which means three boundary surfaces and an 18 dB increase in low-range sound pressure! This is why it's better to stay out of the corners altogether, and near other boundaries it's advisable to use tone control to compensate for the excess low frequencies. Many active monitors have built-in equalizer settings, especially for this purpose. Nevertheless, prevention is always better than cure, because a nearby surface will also create a comb filter: a series of dips in the response that each start from the frequency where the reflected sound arrives exactly out of phase with the direct sound until the frequency where the two sounds are in phase again. These irregularities can't be corrected through tone control. In fact, you'll only make matters worse if you try to iron out the dips electronically (see Figure 17.6).

What you can do to minimize the damage is make sure that different boundary surfaces affect different frequencies. This is why you should place your speakers at different distances from the back wall, side wall, floor and ceiling. The distance to one boundary surface shouldn't be a multiple of the distance to another surface. For example, pay attention that you don't place a speaker 120 cm from the side wall and 60 cm from the back wall. This way, the peaks and dips of different surfaces won't coincide—and thus affect the sound you hear even more.

Acoustic Adjustments

Now that you've decided on the layout of the studio, it's time to take control of the acoustics in the space. It would be pointless for me to give

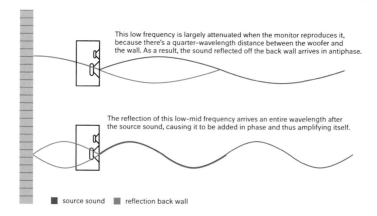

This low frequency is largely attenuated when the monitor reproduces it, because there's a quarter-wavelength distance between the woofer and the wall. As a result, the sound reflected off the back wall arrives in antiphase.

The reflection of this low-mid frequency arrives an entire wavelength after the source sound, causing it to be added in phase and thus amplifying itself.

■ source sound ■ reflection back wall

Figure 17.6 Comb filtering occurs when a direct sound (blue) is combined with a delayed reflection of the same sound (orange). Wavelengths with periods that fit an integer number of times in the delay add up, while wavelengths with periods that fit half a time in the delay—or an integer number of times plus a half—subtract. In practice, the effect gradually becomes less severe as the frequencies get higher, because the speaker only radiates these high tones toward the front.

you general recommendations about the exact measures to take and materials to use (which are different for every studio), but here are some rules of thumb to help you get started:

- It helps to damp the early reflection paths (sound that reaches the listening position after a single reflection) as these reflections arrive at your ears so quickly after the direct sound that they cause an annoying interference (the comb filtering shown in Figure 17.7). It's a good idea to take the absorption that you need anyway to improve the room's frequency response, and position it in such a way that you immediately damp the early reflection paths between the speakers and the listening position, thus killing two birds with one stone. For this application, it's best to use (broadband) absorption that's as thick as practically possible (see Figure 17.8).
- In the rest of the space (provided that it's not too small) you can combine damping and reflection. For example, think of an absorber covered with a pattern of wooden panels. This approach will maximize the part of the space in which the frequency response is uniform. The lack of large surfaces that only damp or reflect creates an even reverberant field in the space.
- Large reflecting surfaces that can't be damped or made diffuse (such as windows and floors) are best compensated by adding a bit more damping and diffusion to the opposite surfaces.

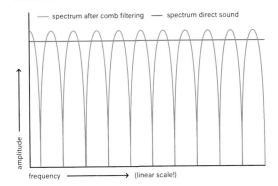

—— spectrum after comb filtering   —— spectrum direct sound

amplitude →

frequency ————————→ (linear scale!)

Figure 17.7 The result is a series of additions and subtractions in the frequency spectrum: a comb filter.

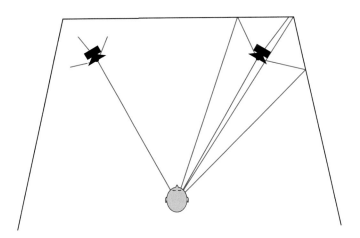

Figure 17.8 Damping the early reflection paths between the speakers and the listening position (left) can significantly reduce the influence of early reflections on the direct sound (right).

- The accumulation of low-range energy in the corners of a space (caused by standing waves) makes corners—and especially those where three surfaces meet—suitable places for bass traps or broadband absorption, both of which reduce the influence of standing waves. Most bass traps are pressure absorbers, so they can only be effective at the position of a pressure maximum (antinode) of the sound wave that needs to be damped (see the box 'Bass Traps'). You can locate the antinodes in a room with a sound pressure meter and a sine generator set to the problem frequency.

- It's hard to go overboard with bass traps, but with the other types of absorption it's definitely possible to use too much! If your studio sounds dull, apart from damping the early reflection paths, it's probably a better idea to use diffusion than absorption.
- Don't put too much carpet on the floor: carpet only absorbs high frequencies, and this loss is difficult to compensate in the rest of the space. It's better to leave the floor reflective, and to compensate for this by installing thicker absorption materials on the ceiling.
- The selection and placement of furniture is important as well. An open bookcase works like a diffuser, as opposed to a glass-door cabinet. Some couches absorb bass frequencies, and thick pleated curtains absorb more midrange frequencies than sheer curtains. Therefore, the furniture is also a collection of acoustic adjustments that you have to take into account, and when it comes to their placement, the same principles apply as to the positioning of acoustic adjustments.

Bass Traps

Sound waves consist of air particles that move toward and away from each other, thus creating high- and low-pressure zones. If you want to absorb these waves through friction, the best place to do this is where the air volume has the highest velocity. These points are located exactly between the pressure maxima, at one-quarter and three-quarters of the wavelength (see Figure 17.9). Considering the fact that there's always a pressure maximum at a wall or any other barrier (so the air volume velocity is zero here), these aren't the best places to restrict air movement. Therefore, absorbent materials are often placed a bit away from the wall and built to be as thick as possible. This is done to maximize the frequency range in which they can have an effect. However, the lowest frequencies would require impractically large distances and thicknesses: a quarter wavelength of 50 Hz is 1.71 meters. So in order to effectively damp these frequencies, you'll need a more realistic solution.

Buffers

When you replace a hard wall with a flexible one, the sound won't be reflected so easily anymore. You can compare it to the difference between dropping a bouncy ball on a concrete floor and on a

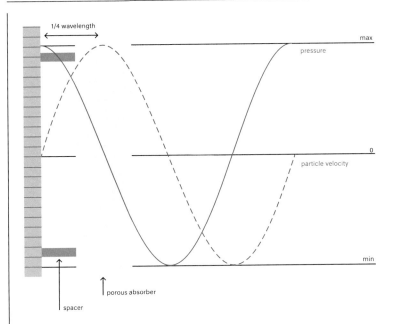

Figure 17.9 A porous absorber can only be effective in places where the air particle velocity is high. This is why these absorbers are often installed with spacers. In order to absorb very low frequencies this way, the distance to the wall and the thickness of the absorber would have to be very great.

trampoline: in the latter case, it won't bounce back nearly as hard, as the trampoline partly absorbs its energy. The problem is that this absorption is frequency-dependent: a very light bouncy ball (a high frequency) will hardly make the surface move and bounces back relatively easily, while a heavy ball (a low frequency) really sinks into the trampoline and is slowed down this way.

A flexible barrier like a trampoline can be seen as a resonator that you can tune to a particular frequency. Damping the resonator enables it to extract energy from the sound and convert it into heat. There are various kinds of tuned absorbers that target low frequencies. Some are really almost like a trampoline, consisting of a spring-suspended membrane covering a closed and damped air chamber.

Helmholtz Principle

Other models use the Helmholtz principle of a vibrating air column in a half-open chamber. A bottle or jar with a narrow neck works

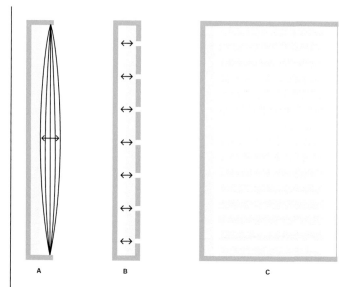

A   B   C

Figure 17.10 Tuned absorbers that use a damped membrane (A) or
Helmholtz construction (B) can absorb a narrow frequency
range very efficiently. They respond to pressure and not to
particle velocity, which is why they are mainly effective against
a wall or in a corner. But if you need to damp a wider range
of frequencies, it's better to install a series of porous
absorbers (C).

like this, but a perforated panel covering a damped air chamber
also produces a Helmholtz absorber. The surface area and depth of
the holes—combined with the volume of the air chamber—determine the frequency that's damped. Unlike porous absorbers, these
tuned absorbers have the greatest effect at high-pressure points,
so against walls and in corners. If you position them in a place
where there aren't any pressure maxima, they can't do their job.

Tune and Spread

The drawback of tuned absorbers is that they have a narrow range
of action. You have to precisely tune them to the problem frequencies in your room, which is not all that easy. Therefore, more
broadband solutions are often a better choice: if you have the space
for it, a one-meter-deep row of porous absorbers can be very
effective, and you can also combine tuned and porous absorbers
into a 'broadband absorber.' Besides these, there are also damped
metal plates on the market with a slightly wider range of action

(VPR absorbers, based on a freely available patent of the Fraunhofer Institute).

The most reliable solution in terms of tuning is the active bass trap. This can be seen as a subwoofer with a filter that's tuned to a specific resonance. When it detects this frequency with its microphone, the woofer generates a sound that's in antiphase with the original, thus attenuating the resonance. This is a very accurate and efficient way to damp a specific resonance, but it would be an impossible task to install a number of these expensive devices for every standing wave. PSI audio has introduced an active solution that has broadband response below 150Hz (thus attenuating all standing waves present at a particular location), making the active route more feasible. Still, the basis of a good low-end reproduction at the listening position remains a space with favorable dimensions and good speaker placement.

Diffusers

Creating good acoustics in your studio doesn't mean eliminating every single reflection by using as much damping as possible. If your room isn't very small, it's better to aim for a reflection pattern that doesn't have major irregularities. Ideally, the reflections don't all arrive at the listening position at the same moment, but are evenly distributed over time. The shape of the space is a big factor in this: if there are fewer surfaces directly opposite each other, the reflections will automatically be more scattered in terms of direction, and therefore in time as well (see Figure 17.11). Besides this, you don't want certain frequencies to bounce around much longer than others. This can happen when curtains largely absorb the high frequencies, but don't have any effect on the low end. And as mentioned before, frequencies with wavelengths that fit an integer number of times between the walls of a space will resonate longer than others. Therefore, the main challenge for a studio designer is to distribute all these reflections as evenly as possible across time and frequency, and diffusion can be very useful for this.

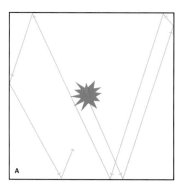
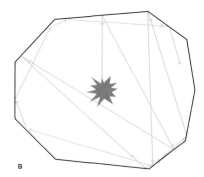

Figure 17.11 In a room with a lot of parallel surfaces, the reflections aren't really spread out in direction and time (A). Room B breaks these regular patterns.

Scattered

Diffusion of sound means that instead of a direct reflection (like a flat mirror), you're spreading reflections in many directions (like a disco ball). You can do this by giving flat walls an irregular surface. The size of the irregularities changes the effect: in a rippling puddle of water you would immediately recognize yourself, but in the waves of the sea your mirror image would turn into something else altogether. The greater the irregularities, the lower the frequency at which they influence the direction of the sound reflection.

A diffuser is an object that contains a lot of different dimensions, designed to spread as many frequencies in as many directions as possible. You can compare it to a puddle of water that has both small ripples and big waves, scattering both the fine details and the main lines of your mirror image. Diffusion is the perfect studio design solution when you don't want to use more absorption (because this would make your space sound too dull or dry), but you still want to reduce the coloration that the reflections add to the sound.

Mathematics

The best diffusers aren't simply a collection of randomly distributed surfaces, rather their design is based on a mathematical sequence. This way, the diffusion character is as balanced as possible. The best known example of such a sequence is the quadratic residue sequence, which forms the basis of the quadratic residue diffuser (QRD). You choose a prime number (for example: 7, 11 or 13) and create a sequence of consecutive integers: from zero up to and including your prime number. Then you square these numbers, and from each of the resulting numbers you subtract the prime number as often as you can. The numbers this produces are the quadratic residues that indicate the depth of the different segments of the diffuser.

You can use this sequence (which you can also find online if you're not into arithmetic) to determine the relative depth ratios of any number of surfaces. And if you want to increase the frequency range, you can also enhance the surfaces of a large diffuser with smaller copies of itself. This way, you're building a fractal diffuser, which resembles the structure of ice crystals. On top of this, you can build two- and three-dimensional diffusers, which scatter the sound across one or two dimensions, respectively (see Figure 17.12). Diffusers (especially if they have enough depth) can greatly improve the sound of a room without it becoming too 'dead.' The only things they can't correct are standing waves in the low end (unless the diffusers are several meters deep). In places where unwanted direct reflections occur (for instance at the side walls, toward the listening position), you're usually better off with absorption.

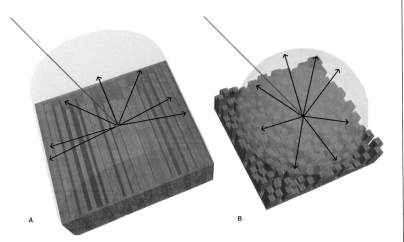

A   B

Figure 17.12 A two-dimensional QRD spreads the sound across one dimension (A), while a three-dimensional skyline diffuser spreads the sound across two dimensions (B).

17.2 Monitoring

At the average music store, you'll easily find dozens of different studio monitor models. So there are plenty of options to choose from, but what requirements should a good monitor meet? If you only look at the most obvious specification, the frequency response, all of these models seem very similar in terms of sound. Of course that's not really the case, but what's the reason behind this?

Radiation

Just like microphones, speakers have directional characteristics. Their frequency response depends on the direction in which the sound radiates: the higher the frequency, the less the sound is radiated to the back and sides, so the duller the speaker becomes in those directions. Besides this, the edges of the speaker enclosure can cause reflections that distort the response (diffraction), and differences in distance between the drivers and the listening position can cause dips in the response due to different arrival times (see section 7.1 for more on this phenomenon). Therefore, if the listener isn't positioned roughly in a straight line in front of the speaker, the frequency response won't be optimal anymore (although some monitors actually have a slightly more balanced sound at a small angle). But what's more important is that the first reflections—which

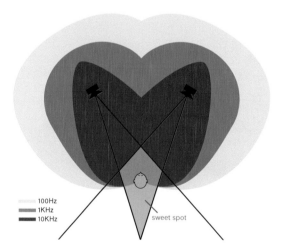

100Hz
1KHz
10KHz

sweet spot

Figure 17.13 Due to the narrow radiation character at 10 kHz, the frequency response projected by these speakers is only balanced in the area between the black lines. As a result, the usable listening position (sweet spot) isn't very large.

blend with the direct sound at the listening position—greatly determine the overall sound. Since these reflections come from a part of the speaker that usually doesn't sound optimal, they can add a lot of coloration to the sound that's heard at the listening position.

It's always better to choose speakers with wide dispersion at high frequencies, as this increases the usable listening position (see Figure 17.13), which will give you more freedom of movement while mixing. But even at the sweet spot in an anechoic room, different speakers with virtually identical frequency characteristics still sound different. The reason for this is that the frequency graphs published by the manufacturers usually don't display the raw measurement data with a big enough resolution (especially in the low end), but instead show a smoothed-out representation of this measurement (see Figure 17.14). The minor irregularities that remain hidden this way sometimes contain a lot of information about the speaker's time response. This is because these irregularities in the frequency response are caused by time effects like resonances of the enclosure and drivers, and filter-induced phase shifts.

Time Response

A speaker has an ideal time response if a short impulse (a test signal that contains all frequencies equally, and which instantly starts and ends) comes out exactly the same way you put it in. This would mean that the speaker has a perfect impulse response, but unfortunately this never happens in real life. The main reason for this is that almost all speakers are multi-way systems with low-frequency woofers and high-frequency tweeters. After the impulse passes a crossover filter, the tweeter reproduces the high part of the impulse and the woofer the low part. It's not easy to precisely match the timing of the woofer and the tweeter, since the filter creates a phase shift that causes the impulse to be 'smeared out' in time (you can read more about this effect in section 17.4). And because the woofer acts as a high-pass filter in the lowest part of its range, it creates an additional phase shift there. In some speakers, the phase response is compensated with inverse all-pass filters, a special filter type that only affects the phase response, not the frequency response. This way, the timing of the woofer and tweeter is aligned for all frequencies, so you'll only hear the problems in the midrange that are actually in your mix, and not the problems caused by the crossover filter.

Besides the fact that all frequencies should be reproduced at the same time, it's also important that they all die out as soon as possible after the input signal ends. You can imagine that if you give a woofer a good push, it will keep wobbling back and forth for a while before it stops. This makes the bass response less tight. In order to control this behavior as much as possible, the amplifier precisely controls the movement of the

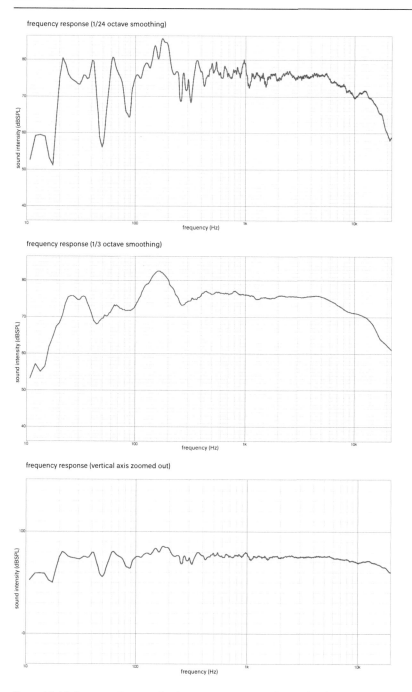

frequency response (1/24 octave smoothing)

frequency response (1/3 octave smoothing)

frequency response (vertical axis zoomed out)

Figure 17.14 By averaging out the frequency response (smoothing) and adjusting the resolution of the vertical axis, you can present the same data much more favorably.

The Numbers Don't Always Tell the Tale

It's not surprising that manufacturers rarely publish radiation characteristics, as the radiation character of a speaker measured in an anechoic testing room says little about the actual everyday listening experience in a studio. There are formulas to convert this data into a prediction of the speaker's performance in an average listening room, but even if you measure the frequency response in your studio, this measurement will only tell you a part of the story. This is because the human ear is much better than a microphone–spectrum analyzer combination at distinguishing between a direct sound and its reflections, and at averaging the minor irregularities caused by these reflections into a more balanced overall sound. Therefore, a single measurement says nothing: the average of a large number of measurements in the area around the listening position would be closer to the real listening experience, especially when it comes to low frequencies.

speaker. It makes the cone move, but it also slows it down again. And the enclosure has an even bigger effect on the speaker's decay. On top of being a small space with its own standing waves that amplify and attenuate certain frequencies, many enclosures also have a bass reflex port. This is a tube with a certain resonance, which helps the woofer to achieve a balanced reproduction of the lowest frequencies. But besides this advantage, bass reflex ports also have many disadvantages. First of all, a resonance automatically means that the low frequencies have a longer decay time than the rest of the frequency spectrum, which has a negative effect on the speaker's time response. On top of this, the tube is tuned at a certain frequency that adds in phase with the woofer at the end of the tube. However, besides the required frequencies there are more frequencies coming from the tube, which might not necessarily be in phase with the woofer and end up being attenuated. And at higher playback levels, the airflow inside the bass reflex port can start to react unpredictably, thus dramatically distorting the speaker's frequency response. A variation of the bass reflex port that counteracts this effect is the transmission line. These tubes are longer and have acoustic absorption, so only the intended frequencies come out of the tube. However, this solution isn't ideal either, as it still involves blending time-delayed copies with the original sound.

For those who don't care so much about the deepest lows, there is a much easier (and better) solution: closed box. This type of speaker might not reproduce the lowest frequencies, but it has a good time response that's free of unwanted resonances. This is the reason why so many professionals still work on Auratones, Yamaha NS-10s or models derived from these. With these speakers, it's easier to make a good assessment of the relationships between instruments, and that's what mixing is mainly about. Many brands make closed-box models, some of which (as opposed to the NS-10) manage to achieve a pretty decent bass response. Just remember that almost the entire consumer world listens to music on speakers with bass reflex ports, so it's a good idea to check your mixes on these speakers every now and then. When choosing monitors, make sure the character of the bass reflex port isn't too prominent. Usually, the monitors that aren't very impressive at first due to their lack of an overwhelming low end prove more reliable for mixing upon closer inspection.

Distortion

The last aspect that's important for the sound differences between speakers is distortion. For example, a woofer only has a limited range in which it can move freely before it's held back by its suspension—extreme excursions will stretch the woofer to its limit, making it less elastic over time. And when you try to reproduce relatively high frequencies with a woofer or tweeter, the weight and lack of stiffness of their membrane can really throw a wrench in things. This kind of distortion occurs in all speakers, and the only thing that helps to keep it within bounds is using high-end drivers and carefully taking the interaction between enclosure, drivers and amplifiers into account. With amplifiers it's the same as with speakers: two different models—both with a frequency response that seems to be drawn with a ruler—can still produce huge differences in sound. A good amplifier is fast and powerful enough to reproduce short peaks (transients) without distortion. This makes it easier to perceive the details in your mix, and to hear if there's too much or too little reverb on an instrument, for instance. A good monitor system won't reach its upper limit as easily as the average consumer speaker. The latter will start to sound 'loud' much sooner when you crank up the volume, due to the compression and distortion it adds. It's for this reason that it takes some time to get used to working on a high-end studio system: you'll have to turn it up much more if you want to evoke the same association with loudness, because it doesn't distort as much. That's why I also have a pair of average-sounding bass reflex speakers in my studio, especially for clients who get insecure about this lack of distortion. As soon as they hear that the mix can produce the familiar feeling of a speaker that's pushed to its limit, they are reassured.

How Do You Choose the Right Monitor?

Few manufacturers publish frequency responses with a big enough resolution to actually say something about the sound of their product, and data on directivity and distortion is even harder to find. So there's not much else to do than to carefully listen to the differences yourself. If you know what to pay attention to, it's a lot easier to recognize problems. Besides this, it's mainly important to build an intimate relationship with your monitors. Once you're familiar with their peculiarities, they don't need to hamper the reliability of your work anymore.

17.3 Subwoofers

A common problem, especially in home studios, is that you don't hear enough lows. The obvious solution would be to install a subwoofer, because if you generate more low end, you're bound to hear it as well. However, this doesn't always solve the problem.

17.3.1 Why Use a Subwoofer?

The idea behind a subwoofer is that it adds something your monitors can't deliver. Therefore, step one is to find out if that's really the case. 'But I can hear right away that there's not enough low end,' you might think. That's undoubtedly the case, but there can be different reasons for this. You can start by looking at the specifications of your monitors. A speaker with a linear (within 3 dB) frequency response until 40 Hz should be able (at least tonally) to play almost all kinds of music in full glory. However, if this doesn't seem to be true in practice, there's a reasonable chance that there are other things going on. Especially the acoustics and the positions of the speakers and listener(s) are crucial to the performance in the low end.

Acoustics can't be changed with a flick of the wrist, but placement is something you can easily experiment with. The basic rules from section 17.1.2 can provide a starting point for this: avoid equal distances to two or more surfaces (walls, floor, ceiling) near the speakers, and don't place them (even when the distance is relatively great) precisely in the middle between two parallel surfaces. Then it's a matter of moving your speakers around, listening and measuring to find the optimal position. Even if you still need an additional subwoofer after you've optimized the placement, it's worth your while to look for the best positions (though you'll soon find out that adding a subwoofer can also affect the optimal placement of the monitors). If your problems can't be solved through placement, there are three possible causes for this, two of which can be solved with a subwoofer:

1. The monitors don't produce enough low end. This problem is easy to fix with a subwoofer.
2. The speakers can't be placed in the optimal position. The location with the best bass reproduction isn't always a realistic option for monitor placement, for example because the rest of the frequency spectrum and the stereo image are out of balance at that position, or because it's impossible due to the layout of the space. This problem can also be solved with a sub, because you can usually place the subwoofer (independently of the monitors) in the optimal position for low-end reproduction.
3. The acoustics are in imbalance. For instance, in a small, square-shaped room with insufficient low-frequency absorption, prominent standing waves make it impossible to achieve a good reproduction of the lows. This is something a subwoofer can't do anything about. Often, what you can do is place the sub in such a way that the main standing waves aren't excited as much, but it will never sound great. Unfortunately, this is a very common situation in home studios.

Does option one or two apply to your situation, possibly with a pinch of option three? Keep reading then . . .

17.3.2 Choosing a Subwoofer

It's not easy to reproduce low frequencies reliably: to achieve a good amount of sound pressure in the low end, you'll have to move a lot of air. In order to do this without audible distortion, you need special resources. Extremely stiff speakers that can endure large excursions before they're stretched too far, amplifiers with a lot of power, and preferably closed enclosures that don't resonate. Besides the fact that you generally get what you pay for, it's important to listen carefully to the subwoofer you're interested in. Properties like distortion and enclosure resonance aren't mentioned by most manufacturers, so it's better to assess these in practice. Two simple things you can check are bass drums (transient response and decay time) and low sine tones (distortion). In a space with good acoustics, you can listen to different bass drums and determine if they sound pointy enough. Do the lows feel tight, or does the bass drum seem to resonate and slowly trail behind the music? Sine tones are handy for detecting differences in sound caused by distortion. Find out if deep sine tones can be reproduced at high levels without audible overtones. The drawback of sine bass is that it strongly depends on your acoustics how loud you hear it. Therefore, while testing a subwoofer, you should walk around the room a bit to get an average impression. If you find it hard to determine what to hear, you can record the sine with an omnidirectional microphone near the woofer and use headphones to compare the recording with the original sine signal.

Welcome Side Effects

Besides the fact that a subwoofer can benefit the frequency response and the interaction with the room, there are a few more advantages. Usually, the low end is the limiting factor for the maximum loudness of your system: the amplifiers will start to clip when you turn things up too high. But many problems—such as speaker distortion—already occur at lower levels as well. By adding a sub, your monitors will need to less work hard, which means they won't distort as much (to take advantage of this, you will need to use the subwoofer with a crossover filter). The reduction of this distortion is much more important than the added potential loudness: the sound will be less tiring to the ears and more open, even at lower levels. To some extent, this openness can be explained by the fact that the woofers don't need to make very large excursions anymore, but it's also caused by a reduction of the pitch modulation that occurs when a woofer tries to reproduce both low and midrange tones. The low tones then start to modulate the midrange tones, because the position of the woofer keeps changing. It's like a violin that's moving back and forth on the frequency of the double bass. You can never eradicate this kind of modulation completely, but by distributing the frequency spectrum across multiple drivers, it will be greatly reduced. This is why three-way systems usually sound much more open and impressive than their two-way counterparts—this advantage has more to do with distortion than with frequency response.

Connections

Besides the sound of a subwoofer, the options it offers are also important. Models with detailed crossover settings are preferred, since they are easier to integrate into a monitoring system. These subwoofers have inputs for the left and right monitor channels, and divide the sound into what they reproduce themselves and what is sent to the monitors. Usually, these are analog crossovers without delay compensation (at most, they might have a phase control, but more on that later).

If you want to go for the ultimate adjustability, it can be a good idea to purchase a separate (digital) speaker processor. This way, you won't need the subwoofer's built-in crossover filter anymore. The advantage of a separate processor is that you can compensate the differences in distance between the monitors and sub(s) with delay, and that it often has its own tone control as well, which you can use to refine the sound. Some subwoofer systems already have such a processor built into them, complete with a monitor controller and digital inputs to keep the number of analog-to-digital and digital-to-analog conversions (and the resulting signal degradation) within limits.

One or Two?

Although subwoofers are usually sold individually and designed for use in a 2–1 setup (one subwoofer with two stereo monitors), it's actually preferable to use two subs. The assumption that you can't hear direction in the deep lows—and therefore that it's no problem to reproduce these in mono—is incorrect. It's hard to hear exactly where a deep bass comes from, but phase differences in the low end can still tell you a lot about the size of the sound source and the acoustics around it. In classical music production (which relies heavily on a realistic spatial experience) it's not done to use high-pass filters casually, because even if the lowest note of a string quartet stays above 80 Hz, you'll still feel the background noise of the hall, deep below the music. This greatly contributes to the feeling of being in the same space as the musicians.

Besides the fact that stereo lows can sound better and more realistic than a mono reproduction, there are more reasons to use two subwoofers. First of all, it's much easier to align them (in time/distance) with the main monitors, thus reducing the problems in the crossover range (see Figure 17.15). On top of this, two subs average out the influence of standing waves by exciting them in two different places. This can help to make the lows sound more even.

17.3.3 Placement and Tuning

When looking for the best place for your subwoofer, you need to balance two things: its interaction with the monitors (I assume their placement is already optimal) and with the acoustics. If you use a single subwoofer, you want to make sure that the monitor–subwoofer distance is equal for both monitors. Only if this causes a serious acoustic problem (which is not unlikely, as the sub often ends up exactly in the middle of the room if the distance to the monitors has to be equal—another advantage of using stereo subs) you can opt for an asymmetrical placement for the sub. Is delay compensation not an option for you? Then it's best to place the sub at the same distance (taking the height difference into account) to the listening position as the monitors. You'll notice that this severely limits your options in terms of placement: only a small part of the imaginary

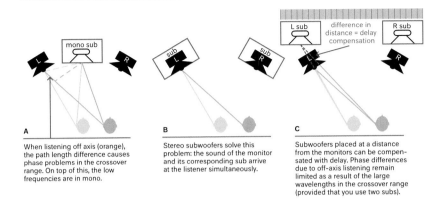

When listening off axis (orange), the path length difference causes phase problems in the crossover range. On top of this, the low frequencies are in mono.

Stereo subwoofers solve this problem: the sound of the monitor and its corresponding sub arrive at the listener simultaneously.

Subwoofers placed at a distance from the monitors can be compensated with delay. Phase differences due to off-axis listening remain limited as a result of the large wavelengths in the crossover range (provided that you use two subs).

Figure 17.15 When listening outside the sweet spot, two subwoofers (B) are favorable compared to a mono sub (A), as there's no difference in distance, and therefore no phase difference in the crossover range. To a large extent, this advantage still applies to situation C, for although there is a difference in distance here, it's very small compared to the wavelengths in the crossover range.

circle on which your monitors are placed can be used to position the sub. This way, the interaction with the monitors might be optimal, but acoustically it's not necessarily the best place for the sub.

For this reason (having enough placement options, in order to make the best use of the acoustics) it's good to have a digital crossover with delay compensation. This means that you can apply a completely new placement strategy, which makes the most of both the sub and the monitor speaker, and which limits the interference of nearby surfaces. This kind of boundary interference can never be avoided entirely, but you can make sure that its negative effects (mainly a deep dip in the frequency response) take place beyond the range of the speaker in question. After all, the frequency of the dip depends on the distance between the speaker and the boundary surface: the nearer the surface, the higher the frequency of the dip. Therefore, if you place a subwoofer on the floor and close to the back wall, the dip is so high up the spectrum that the subwoofer won't reproduce this frequency anymore. However, the low end will be boosted considerably due to the in-phase reflection, but this boost is relatively even and can be compensated by turning down the sub a bit, or by using EQ. On top of this, you can place the monitors far away from the wall, making the first interference dip so low in frequency that the subwoofer has already taken over from the monitors, preventing the dip from happening. Optimization based on this principle is therefore an interaction of the right placement and crossover frequency (see Figures 17.16

and 17.17), which is why it's good to have a crossover with an adjustable frequency. Often, it's not the frequency range of your monitors but the acoustics that determine the optimal crossover frequency.

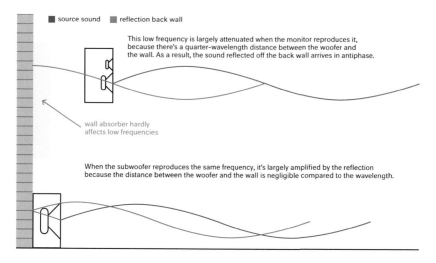

Figure 17.16 Placing a subwoofer against the wall(s) improves the reproduction of low frequencies.

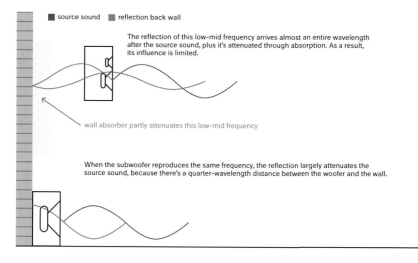

Figure 17.17 Placing monitors further away from the wall(s) improves the reproduction of higher frequencies. In both situations, a less reflective wall means more flexibility in terms of speaker placement, which is why it's always a good idea to install efficient bass absorbers (such as VPRs) on the wall behind the speakers.

Sometimes you can have bad luck, when a subwoofer is placed right at the point where a standing wave has a pressure maximum. As a result, the subwoofer will excite this standing wave over and over again, which will make its frequency resonate strongly. You can never get rid of this resonance completely, but what you can do is try to move the sub to an area where the standing wave doesn't have a pressure maximum, so the subwoofer won't excite it as much.

Fine-Tuning

Things will get easier once you've found the best place for your subwoofer, though you sometimes find out during tuning that you need to reconsider the placement. The main thing to check is the polarity of the subwoofer compared to the main monitors: do all the cones move forward at the same time, or does the sub move backward while the monitors go forward? If they're placed at equal distances, you'll immediately hear a gap in the crossover range if the polarity is reversed. But if you chose a different position for the sub (which you didn't compensate with delay), it can be less clearly audible. Polarity is easy to set: there are only two options, so you simply pick the one that produces the fullest low end. Some subwoofers have a stepless phase control, which is nothing but an all-pass filter that slows down the lows a bit around a certain frequency. You can see it as an analog delay, and it can help to put the lows of the sub and the monitors in phase if you don't have delay compensation.

I should add that such a control can only slow down the lows of the subwoofer, while in practice it's usually the sub that's too late, since it's placed further away. Phase control is therefore not a good substitute for delay compensation, but it can help to make the crossover range—where the subwoofer interacts with the monitors, so where phase is important—more even. Even if the sub is too late, it can still be in phase with the monitors at the crossover frequency. If there's exactly one entire wavelength between the two signals (or a polarity change plus half a wavelength), the result might not sound as tight as a time-aligned system, but at least there are no (prolonged) dips in the frequency spectrum (see Figure 17.18).

In order to achieve the best possible phase alignment of the sub and monitors in the crossover range, you can play a sine wave of the exact crossover frequency. Switch the subwoofer's polarity and turn the phase control of the sub (or the delay compensation of the crossover) until the tone is at its softest point. It's very important that you're exactly at the sweet spot when you listen to the result. If you feel that you've reached the maximum subtraction between the sub and monitors, you can sometimes still increase the attenuation by adjusting the level of the sub. If you then reverse the polarity of the sub again, you can be sure that it's

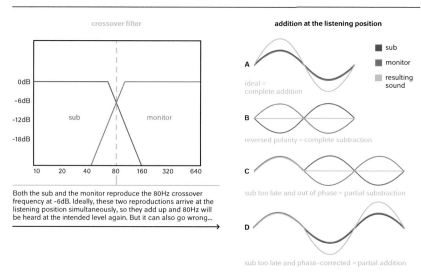

Both the sub and the monitor reproduce the 80Hz crossover frequency at -6dB. Ideally, these two reproductions arrive at the listening position simultaneously, so they add up and 80Hz will be heard at the intended level again. But it can also go wrong...

Figure 17.18 Possible scenarios for the addition of the sub and monitor signals in the crossover range. The ideal situation would be A, since the sub and monitor arrive at the listener simultaneously, and therefore in phase. The worst situation is B, which is identical to A, but with the sub and monitor having opposite polarities. As a result of this complete subtraction, the entire crossover range is attenuated. A common scenario is C, in which a difference in distance creates a phase difference in the crossover range, so in the worst case this range is completely attenuated. This can be compensated for with a polarity switch and fine-tuned with a stepless phase control. As a result, the sub might arrive too late, but when it does, at least it's in phase with the monitor (D).

Subwoofer or LFE?

A subwoofer is not the same as an LFE speaker. LFE stands for low-frequency effects channel, and it's used to reproduce low-range peaks in movies (in music, its use is very limited). The main difference with a subwoofer is that an LFE doesn't reproduce any sound from the other channels, so the channel isn't meant to take over the lows from your main monitors. The confusing thing is that in practice, the LFE speaker (both in small studios and in living rooms) often serves as a subwoofer as well. Since the other speakers can't reproduce the lowest frequencies, manufacturers rely on a bass management system to filter the lows from the main channels and send them to the subwoofer, but then the LFE channel is still added to this.

What can you do with this information? Not so much actually, but there's one major pitfall. If you use an LFE channel as a subwoofer while mixing (for example, by sending the lows from the left and right channels to the LFE as well), chances are that the bass management in the listener's living room will repeat the same process. It's very doubtful that the lows that are sent to the subwoofer by these two routes (LFE and bass management) will actually arrive there in phase—resulting in all kinds of unpredictable additions and subtractions. So if you decide to use an LFE, only use it for its intended purpose: as a discrete channel for huge explosions and other sound effects.

To prepare itself for these big bursts, an LFE uses a calibrated system that's set ten decibels louder than the other channels. This gives your system extra headroom, so it can cope with strong peaks without clipping. However, if you also want to use the LFE channel

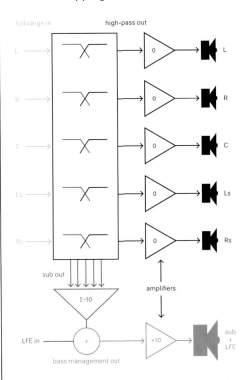

Figure 17.19 How bass management works in a 5.1 system. The signals for the subwoofer and LFE are combined, but first the subwoofer's signal has to be attenuated by 10 dB, since the LFE's amplifier is usually set to +10 dB.

as a subwoofer (with bass management), it's important to make
sure that the sound levels are correct: the lows that are sent to
the LFE with bass management should have the same level when
produced by the sub and by the main monitors (nominal level),
while the sound you send to the LFE should come out at +10 dB.

as much as possible in phase with the monitors in the crossover range.
The final step in this process is to set the level of the sub compared to the
monitors, which you can do by ear. Listen to a lot of different material
that you're very familiar with, and judge if the low end is too big or too
modest. Once you've found a setting that works well, try to get used to
it for a while, otherwise you'll keep searching. Your frame of reference
should have time to adjust to the new situation.

Measuring

If you can't find a setting that feels good, or if some reference tracks
sound perfect and others out of balance, then more thorough research is
required. An omnidirectional (measurement) microphone, a transformer-
less preamp and a software package can provide the solution in this case.
There's a reason why I've waited so long to mention these tools, as they
can also hugely distract you from what you hear if you don't know what
to pay attention to. First of all, it's important to conduct multiple
measurements (a dozen or so, within a one-meter radius around the sweet
spot) to get a realistic impression of the sound. As a listener, your head
isn't stuck in a vise at the exact same spot, so the average sound at and
around the listening position is decisive.

Besides this, you shouldn't focus too much on the details of the
frequency response: minor irregularities are normal, it's only the excesses
that are interesting. The decay time per frequency says much more than
the frequency response alone. A great way to visualize this data is with
a waterfall plot, as this will make standing-wave resonances and
interference gaps stand out much more (see Figure 17.20). These kinds
of measurements can help you to determine the cause of a problem. If
you position the subwoofer and the measurement microphone in different
places and conduct multiple measurements, you'll learn a lot about how
your room works. Big dips that show up at every microphone position
point to boundary interference. You can verify this by placing the
subwoofer closer to or further away from the suspected cause. This will
change the frequency of the dip—that is, if you were right. This type of
interference can be minimized by following the method shown in Figures

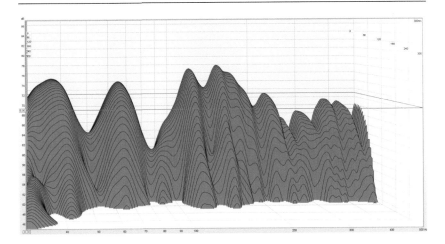

Figure 17.20 Waterfall plot.

17.16 and 17.17. Dips and peaks that do change when you reposition the microphone are usually caused by standing waves. Moving the subwoofer or your listening position can make all the difference—sometimes a meter is all it takes. But if your room is really giving you a hard time, you're fighting a losing battle.

The final step (besides moving) is to reduce the effects of these unsolvable problems as much as possible. With EQ as the last resort, you can still correct the excesses somewhat. Most speaker processors have built-in equalizers, and sometimes monitors and subwoofers also have basic correction filters. Still, you should only use EQ to slightly attenuate the most annoying resonance(s), not to fill gaps. These gaps are caused by subtraction, and if you fill the room with more sound energy at the frequency that's already subtracted, the subtraction itself will become stronger as well. This is a waste of energy, and it can lead to a system overload.

17.4 Headphones

When you're setting up a studio, the acoustics are by far the biggest challenge. If you have little luck with the space that's available to you, fixing it up can quickly become a costly undertaking. Cutting these costs is not an option, as you've already read enough about the adverse effects this can have on your mixes. Therefore, it's not such a crazy idea to sidestep the problem of acoustics altogether with a pair of high-end headphones. No more standing waves that throw the lows off balance, no more reflections that add coloration to the midrange—all you hear is your mix and nothing else. On top of this, it gives you a lot of freedom:

you can take your reference with you, wherever you go. And since you're much less of a noise nuisance now, you can still pop over to the neighbors for a cup of sugar. Nothing but benefits, right? Then why do you see so few professionals working on headphones?

Masking

I once heard someone say about his high-end headphones that he could hear too much of the music on them. Now that he was able to make out every single detail in the mix, it felt like the recipe for a delicious soup was revealed, causing the separate ingredients to stand out more than the whole. Other people tell me they enjoy how headphones allow them to perceive all the musical details that are lost when they listen in their car. Whatever you prefer, you're bound to hear more on headphones than on the best speakers. This is what makes it so difficult to mix on headphones: it's much easier for things to sound acceptable. On speakers, one decibel of EQ can make a big difference to the way in which instruments combine in the mix. But on headphones, these corrections can seem insignificant, because you can still make out all the instruments anyway.

The ease with which you can hear things on headphones is caused by a lack of masking. You can imagine that in a typical listening space, reflections can overshadow details. The shortest reflections are the most powerful, and due to phase interaction they affect the timbre the most. In a two-speaker setup, the very shortest reflection is actually not a reflection, but the sound of the right speaker arriving at your left ear and vice versa. As a result of this 'cross-pollination,' even in an acoustically dead space there will be more masking with speakers than on headphones. Which is not necessarily a bad thing—it's actually the very reason why stereo works so convincingly. The two speakers project a sound field that appears to stretch out in the distance in front of you, while on headphones all of this seems to take place inside your head. This mechanism, which is called crosstalk, can of course be easily simulated on headphones by blending a bit of the left channel (with a slight delay, and maybe some EQ) with the right, and vice versa. Unfortunately, we can't be fooled so easily: due to the fact that the added delay doesn't change when you move your head, it creates a very static filter effect, which you would never experience with speakers.

Tips for a Successful Mix

Low Frequencies

Sound waves that travel through the air not only enter your body through your ear canals, but low frequencies can also pass through soft barriers like skin and muscles. This is one of the reasons why your perception of the deepest part of your mix is completely different on speakers than on headphones, even though the sound pressure level at the ear is the same. Therefore, it's not very smart (and potentially even dangerous) to try to achieve a speaker-like impact on your headphones. You can crank up the

Speakers on Your Headphones?

Working on headphones would be a lot easier if they sounded like a perfect speaker setup in a perfect studio. Therefore, it's no surprise that there are a lot of products, both software- and hardware-based, that try to achieve this. The fun thing is that you can often bypass their processing while listening, so you can immediately hear how your mix sounds on normal headphones. It's the effect of switching between speakers and headphones, but all on the same headphones.

The simpler designs are based on a form of filtered crosstalk. This creates a more realistic panorama and adds more masking to the mix, but it still doesn't sound like speakers in a room. More advanced products use impulse responses of real studio speaker setups recorded with a dummy head (with two microphones in the auricles) to process your audio. This way, you'll hear more 'air' around the sound, which closely resembles the effect of listening on speakers.

However, these systems don't really work either. The point is that they don't react to head movements. The directions where the reflections and the direct sound come from stay constant, no matter how you turn your head. Therefore, the result still sounds like a clearly identifiable static filter, while the filter effects in a real space constantly change with the movement of your head—and attract much less attention. This makes direct and reflected sound easier to distinguish in a real space, plus the spectrum will sound more balanced. The only systems that come close to real speakers are those that use head tracking to adjust their processing to the position of your head (which takes quite a lot of processing power, as you can imagine). Personally, I feel that speakers-on-headphones solutions make equalization much more difficult, because they add a fair amount of coloration to the sound. For a check during mixing I do find them interesting, as they can tell you if have enough definition in your mix to pass the speaker test.

volume as much as you want, but you'll never feel the bass in your guts. The only thing you can do about this is familiarize yourself with how the low end is supposed to sound on your headphones, and how it then comes across on speakers. Listening to reference tracks a lot and comparing these with your own mix can certainly help.

Mono

On headphones, wide panning works even better than on speakers if you want to make instruments stand out in the mix. Which is not surprising, considering the fact that the sound field between two speakers isn't much wider than 60 degrees, while on headphones it's stretched to 180 degrees. To prevent wide-panned instruments from becoming so soft in the mix that they're drowned out on speakers, you can listen in mono every now and then. If all the instruments are still audible then, you can assume you'll still be able to hear them on speakers as well.

Go Easy on the Reverb

When you play your mix through speakers, the listening space adds its own reverb. This is why the acoustics in good studios have a controlled amount of reverberation, instead of a bone-dry sound: to give you an idea of how the mix will sound in a living room. But headphones can't provide this idea at all, of course. Therefore, it's better to err on the side of caution when adding reverb, especially with short ambience reverb that resembles the reverb in an average listening space. You need to keep some room for this 'living room reverb' to be added later if you don't want your mix to become woolly. Sometimes you really have to restrain yourself with reverb, because on headphones you can go much further before the direct sound is drowned out. This is because it's much easier on headphones to locate sound sources in a stereo image and to isolate them from the surrounding sound. In mono, this effect is gone, so if you really have to use headphones to assess the right amount of reverb, the mono button is your friend. Longer echoes or reverbs with a fair amount of pre-delay are easier to set on headphones. This is because there's relatively little overlap with the source sound—and with the acoustics of many listening spaces—thus reducing the amount of masking.

Dotting the i's

If you're going to work on good monitors for the first time, you have to get used to the fact that your mix can already sound pretty good, while it's still far from finished. You'll learn to slowly raise the bar and try to achieve even more definition in your mix. Otherwise the result will sound

Types of Headphones

Headphones can be made for a number of specialties: isolation, portability/comfort, visual inconspicuousness or audio performance. Often, these specialties don't go together in the same design without making some compromises. For example, headphones that offer extremely good outside isolation often have to make concessions to sound quality. The closed chambers required for the isolation reflect the sound coming from the back of the drivers. And in-ear designs can definitely sound good, but they're extremely sensitive to minor variations in how they fit in the ear canal, which dramatically affects their low-frequency response. On top of this, they sound even less like speakers than over-ear headphones, as they completely eliminate the influence of your auricles.

Headphones that are purely built for audio performance are of course preferred for mixing. Often these are open-back models with little isolation and a relatively large driver close to the ear (circumaural). These models don't squash your ears like supra-aural headphones do, which benefits both the sound quality and the wearing comfort. The relatively large drivers make it easier for open-back headphones to reproduce low frequencies, but they do ask more of your headphone amplifier: you can't expect the output of your laptop to control the drivers as well as a dedicated design with more power and better components. Still, a decent headphone amplifier and a pair of high-end headphones can be purchased for less than 500 dollars, while studio monitors in this price range are usually not exactly the cream of the crop.

In an even higher price category, you'll find exotic headphones that use electrostatic drivers instead of moving-coil speakers. An electrostatic element consists of a thin membrane that's electrically charged. It's suspended between two conductors that generate an electromagnetic field based on the incoming audio signal. This causes the membrane to move. Because it's so light, the high-frequency response is very good and the harmonic distortion very low. The drawback is that it takes hundreds of volts for the system to work, which makes this type of headphones less suitable for use on the road. But whatever type you choose, it's advisable to first listen carefully if a particular pair of headphones does what you expect of it.

like a woolly mess on average speakers, with the instruments getting in each other's way. On headphones, the bar should be even higher. Therefore, my main advice is to be extra critical of your mix balance. Even if everything is already in its right place, you can still try to hear which frequency ranges are dominant. Which elements draw too much attention? Are all the important elements sufficiently emphasized? A finished mix will usually feel overdefined on headphones.

Final Check

A final evaluation of your mix on speakers is always a good idea. Of course, you can outsource this task to a mastering engineer, who will usually point it out to you if something really doesn't work in your headphones-only mix. But a listening round on the various speaker systems you come across can also give you a good impression of the things that can still be improved. And once you've figured this out, you can usually hear these problems on your headphones as well, and then use your new insights to fix them.

If you gain some experience with how headphones work, they can be a great reference, especially when you combine them with speakers. That's also how many professionals use them: they don't want to run the risk of consumers hearing weird things on their headphones that they had missed on their speakers.

17.5 Equalizers and Their Side Effects

Technology: Phase

You can use an equalizer without knowing how it works on the inside, but it's still good to know that it changes more things than just the frequency character of a sound. Equally important for the sound of an equalizer is how it behaves in time, because all analog and most digital (infinite impulse response, or IIR) equalizers create a phase shift. This is a frequency-dependent delay, which causes the different frequency components of a sound to shift in time in relation to each other. The sound is 'smeared,' which can cause its impact to wane and its details like depth and stereo placement to fade. For example, think of a bass drum of which you first hear the 'click' of the attack and then the low-end rumble. It will sound less tight and punchy than a kick with all the frequency components making their entry at the same time.

In many cases, phase shift is not a problem, and it can even sound very natural. The filter effects you perceive in everyday life (caused by acoustic barriers) also create phase shifts, so you're very familiar with the sound. It only becomes a problem when the shifts are large and/or numerous.

The amount of phase shift caused by an equalizer depends on the amount of amplification or attenuation of a particular frequency band, and on the bandwidth. In addition, every extra band that you use on the equalizer causes its own phase shift.

Tip: Less Is More

All the negative effects of EQ will be greater if you use more cut or boost, and a higher number of filter bands also makes things worse. Therefore it's important to equalize with a clear vision. For example, you can try to balance a honky-sounding piano with a narrow boost at 80 Hz, a wide boost at 4 kHz and a shelving boost at 12 kHz. But you could also choose to make a little dip at 250 Hz and turn the piano up a bit. Perceptually the effect is almost the same, but because you've only used one filter instead of three, the piano will sound richer.

Always try to avoid having a filter in one place in the signal chain compensating for another filter somewhere else in the chain. This sounds obvious, but many people listen to individual elements on solo when they equalize them, and then compensate for this contextless channel EQ with an additional equalizer on the mix bus. Though it's true that EQ sounds different on the mix bus, it should always be a conscious musical choice, and not a compensation for poorly chosen channel EQs. Before you know it, you've lost all depth, fullness and detail.

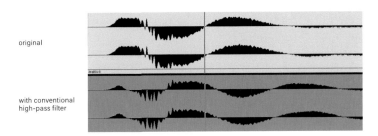

Figure 17.21 A filtered bass drum changes in time. In the processed waveform, the long sine shape is shifted backward, as the low frequencies slow down more than the high frequencies.

Technology: Phase-Linear Filters

Digitally, it's possible to build phase-linear (finite impulse response) equalizers, which—you've guessed it already—don't cause phase shifts. This makes them ideal for adjusting your frequency balance without affecting the impact and subjective timbre of the music. For the post-processing of classical recordings and in mastering, this type of equalizer

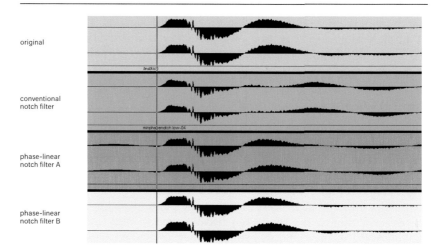

Figure 17.22 A very steep 110 Hz notch filter applied to the bass drum from Figure 17.21, using a conventional filter and two different phase-linear filters. The conventional filter has no pre-echo, but a lot of phase distortion. Phase-linear filter A has a substantial pre-echo, while phase-linear filter B hardly produces any pre-echo, thanks to its different design. Therefore, phase-linear filter B is probably the best choice for this particular notch filter.

is popular as a troubleshooter. For example, if the tiniest adjustment with a conventional EQ already makes your stereo image collapse, phase-linear EQ can offer the solution.

However, these equalizers have an Achilles' heel as well: they cause a minuscule echo before (pre-echo) and after the sound they process, due to the fact that their impulse response is symmetrical. Because our ears are so focused on signaling the first arriving sound, the pre-echoes in particular can make a sound feel unnatural. Therefore, phase-linear EQ is definitely not the best choice in all cases. Especially the treatment of low frequencies with steep filters is a challenge, as this leads to longer and more intense echoes. A bass drum can then sound like a 'flob' instead of a 'pock.'

Tip: Wide Is Better

Filters sound most natural when their bandwidth isn't too narrow. Or in the case of high- or low-pass filters: when they aren't too steep. Phase shifts and other side effects are less extreme at a greater bandwidth, and phase-linear equalizers are less affected by pre-echo and other side effects at a greater bandwidth. Therefore, if no major surgery is required, try to limit yourself to coloring with broad strokes.

Technology: Resonance

Another way to look at the temporal behavior of equalizers is by focusing on the rise and fall time of their resonance. Analog filters use the resonance of a coil (passive), or feedback around an amplifier (active) to achieve the desired frequency response. Conventional digital filters use algorithms that form the equivalent of those circuits. As a result, they all tend to resonate around their operating frequency (ringing). It takes some time for the filter to reach its optimal effect, and it decays for a while as well. This resonance also causes the filter to slightly increase the signal level around its operating frequency—this is called overshoot. The narrower

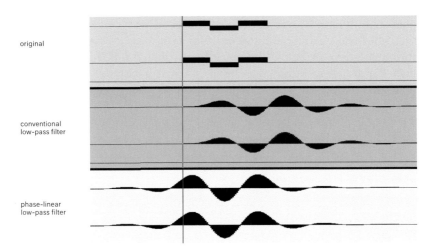

original

conventional
low-pass filter

phase-linear
low-pass filter

Figure 17.23 A I kHz square wave filtered with two steep low-pass filters, with the cutoff frequency at I kHz. A theoretically perfect filter should allow a I kHz sine with a length of one and a half periods to pass. A number of things stand out:

1. The amplitude of the filtered signals is much higher than that of the original. This is because the filter resonates at the same frequency as the fundamental of the source signal. As a result of this resonance, a signal can have higher peaks after filtering than before, even if the filtering should theoretically reduce the signal's energy. This phenomenon is called overshoot, and it's a form of distortion.
2. The conventionally filtered signal only reaches its maximum amplitude after the first period, when the resonance of the filter is added in phase with the direct signal and overshoot occurs. This disruption of the amplitude ratio between the first and second period is also distortion, and it shows that EQ can distort transients—and therefore the definition of a sound.
3. The amount of resonance in both filters is visually clear from the long echo.
4. The amount of pre-echo in the phase-linear filter is considerable, and listeners will probably perceive it as disturbing.

the bandwidth of the filter, the longer and stronger the resonance. With some filters you can set the bandwidth so narrow that you can use them as a permanently resonating oscillator.

The resonance of an equalizer can seriously ruin a sound. Before you know it, you're hearing hissing, whistling, ringing or buzzing noises. The lows in your mix can become less tight, the midrange more aggressive and the highs start to whistle if you use too narrow filters.

Tip: Create an Illusion

Often, you need to limit the frequency range of individual sounds to make them fit into the mix. Low- and high-pass filters are useful tools for this, but it's not easy to gain enough room without the sound becoming unnaturally filtered. Counterintuitively, a good way to avoid this is by using filters that actually add more resonance. This creates an amplification just above or below the range that's filtered out. As a result, it's much less noticeable that the sound has lost a part of its frequency range—in a way, the amplification masks the filtering. Sweeping the frequency of a resonant high-pass filter can almost feel like you're tuning an instrument after the fact: it gives a sound a new center of gravity that resembles the resonance of a soundbox.

Technology: Upsampling

Digital equalizers have a problem their analog counterparts don't have: the sampling rate you use restricts their frequency range. Above half the sampling rate (the Nyquist frequency) a digital system can't code data anymore, so the higher the sampling rate, the more natural shelf and band filter curves can be on high frequencies. This is why many digital equalizers internally convert the signal into a higher sampling rate before they perform their manipulations. After this, they convert the signal back to the original sampling rate using a phase-linear downsampling filter. As transparently as possible, this filter removes the ultra-high data (which can't be coded at the original sampling rate) from the signal, to prevent aliasing. This entire process is called upsampling, and the implementation of it is very important for the sound (and one of the reasons why not all digital equalizers sound the same).

If the quality of the downsampling filter isn't optimal (to save processing power, for example), a mix that has 24 channels processed with this EQ will sound like you ran it through a cheap AD converter. This is a great advantage of working at high sampling rates: the downsampling filter can do its work far above the audible range.

Technology: Harmonic Distortion

It should be clear by now that a filter does a lot more to the sound than you would expect at first. But besides noises you would rather do without—like the 'restless sound' caused by pre-echoes or upsampling—there are also effects that can add something desirable to the signal. The best known example of this is harmonic distortion in analog equalizers. An analog circuit doesn't perform according to an ideal model: analog components have their limits, and their response is often only partially predictable (linear), especially when they reach the edges of their range. The resulting distortion can smooth out the filter's overshoots at extreme settings, or cause the same boost with an analog EQ to have a perceptually greater effect than its digital equivalent.

Theoretically, these features are the only difference between analog and digital IIR filters with the same internal architecture. Their frequency and phase response is the same, so the distortion character is all that remains to explain the difference. This also shows why sampling equalizers or hardware clone plugins don't sound the same as the original. These applications try to emulate existing hardware by applying impulse responses of the hardware to the signal (this principle is called convolution), but harmonic distortion can't be captured in a single impulse response. Still, there are quite a few digital equalizers on the market today that add distortion to the signal to create these analog-like side effects.

Tip: Many Hands Make Light Work

Differently designed equalizers all have their strengths and weaknesses, so it's important to employ them in the field they excel in. You can extend this advice to the treatment of an individual element in a mix with a sequence of different equalizers, all of which contribute a tiny bit to the overall correction. But how can you reconcile this philosophy with the idea of 'less is more?' The system works when each equalizer only deals with a specific problem and has an otherwise neutral setting. Because you limit their use to the problems they excel at, the EQs only need to make minimal corrections, thus keeping the signal as intact as possible.

An example: a vocal track sounds thin, far away and very dynamic, but it tends to become 'hissy,' plus there's an annoying 500 Hz acoustic resonance in the recording. How do you solve this? First you use a phase-linear filter to curb the resonance. 500 Hz is in the middle of the singer's tonal range, so a phase shift caused by a narrow conventional filter would be very noticeable, and make the voice sound even thinner than it already did. Next you exaggerate the low end with a wide passive EQ, which will make the voice seem much bigger. Then there's a compressor in the signal path that mainly responds to the now slightly exaggerated lows, and less to the range that sounded thin and shrill. Now the voice sounds full and its dynamics are in balance, but it still needs some extra highs to make it seem closer. For this, you need an equalizer with a minimal tendency to resonate, as this would only exaggerate the hissiness. So the final step is a wide boost in the highest frequency range, with an EQ that uses little feedback. Problem solved. It should be clear why the equipment racks (or plugin collections) in mixing and mastering studios sometimes reach to the ceiling: good luck pulling this off with only one EQ.

17.6 Compressors and Their Side Effects

Whether you're working in a top-of-the-line analog studio or scouring the internet for plugins: when it comes to compressors, you won't be at a loss for variety. But why are there so many different models? Wouldn't you rather work with twenty units of one very good compressor? For hardware, the answer is: no, there's not a single compressor that can perform every task equally well. In the case of software, it could be theoretically possible, but the complexity of this all-encompassing super-compressor would probably make you dizzy. This is because sound engineers want more from a compressor than just a perfectly unobtrusive attenuation of loud passages. For them, utilizing the sound of compression has become an art in itself. Besides as a tool for moderating dynamics, you can also use compressors to make sounds woollier or clearer, or to give them more attack or sustain. You can emphasize or temper their rhythm, and even slow down or speed up their feel. With different types of distortion, you can make sounds milder or more aggressive. Suggest loudness, highlight details, bring sounds to the front or push them to the background—the possibilities are endless.

You could say that the side effects of a compressor are just as important as the actual compression. These side effects make different compressors excel in different areas. With the same amount of compression, compressor A can make a singer sound nasal and pushy, while compressor B gives her a mild sound. To use these characteristics efficiently in your mix, it helps to know a little bit more about different types of compressors and how they work.

Forward or Backward

A compressor measures the loudness of a signal, and based on this measurement it decides whether or not to attenuate the signal. The measurement can take place at two points in the signal path: before or after its own influence (see Figure 17.24). If it measures the signal at the input, it acts as a feedforward compressor. In this system, the amount of compression only depends on the input signal. If it measures the signal

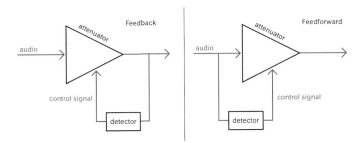

Figure 17.24 The difference between feedback and feedforward compressors.

at the output, it acts as a feedback compressor. The amount of compression then depends on the input signal and the amount of compression already taking place.

Both approaches sound completely different. In a way, a feedforward compressor says: 'The signal exceeds my threshold by an *x* amount, so I lower the signal level by a *y* amount. Period. Next signal.' A feedback compressor, on the other hand, evaluates its own work and says: 'This signal is getting too loud, so I'll turn it down a bit. Not enough? A bit more then. Okay, still some more. Yes, that's more like it.' This means that, fundamentally, a feedforward compressor has a hard knee, and a feedback compressor always has a soft knee (see Figure 17.25).

On top of this, the attack and release times of a feedback compressor are always partly dependent on the input signal (program-dependent), while in a feedforward compressor this isn't necessarily the case. And finally, a feedback design can have a maximum ratio of 5:1, while a feedforward design allows for larger ratios.

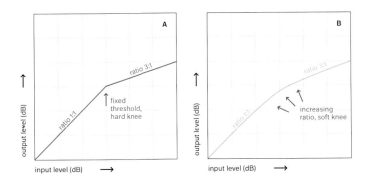

Figure 17.25 Compressor A has a hard knee, so it will only start working above a fixed threshold level. Compressor B is set at the same ratio, but has a soft knee. As a result, it doesn't have a clearly definable threshold, and it will already start working at a much lower level than compressor A.

In the old days, almost all compressors used feedback, and for a good reason. A feedback compressor is self-correcting in a way, as the detector measures the result of the compression it produces. Therefore, in a feedback compressor, the requirements placed on the attenuator's response are less high than in a feedforward compressor. Components that don't have a completely linear response—such as tubes or optical cells—are corrected by the feedback. Only when DBX introduced the VCA (see below), feedforward designs became possible as well. In terms of sound, this offered new possibilities thanks to the hard knee, high compression ratios and tightly fixed attack and release times. But it doesn't end there: the control signal can be manipulated, for instance to still allow for signal-dependent attack and release times, or to change the shape of the knee. A feedforward compressor can therefore be a chameleon, its sound fully dependent on the design of the detection circuit.

Conversely, the response of a feedback compressor can also be manipulated, but it will still sound different. Generally speaking, feedback has a more 'fluid' sound, a highly signal-dependent response and a seemingly milder form of peak reduction. Thanks to their complex response, feedback compressors are often perceived as 'musical.' Although feedforward compressors can also have a fluid and mild response, they distinguish themselves by their potentially extreme peak limiting, static response in terms of timing, and minutely adjustable threshold and knee. For example, their static timing makes feedforward compressors very suitable for sidechain compression of the kind that's used a lot in EDM, or for adding a rhythmically 'pumping' movement to the music.

Besides for peak limiting, the combination of a hard knee and a fast static release time can also be used to introduce distortion and other extreme compression effects. Today, there are quite a lot of compressors on the market (both hardware and software) that can be switched from feedback to feedforward. This is a great option, which really gives you two completely different characters in one device. However, keep in mind that this switch also affects all the other settings (attack, release, threshold, ratio and knee)!

On an Old Fiddle . . .

Compressors have been around for more than seventy years, and during that time various technologies have been invented, which all work a bit differently. The limitations and peculiarities of some of these technologies have sometimes led to musically very useful side effects. Many of the classic designs—or their derivatives—are still in professional use today. And since most of the compression plugins are also based on the operation and sound of old designs, I'll review the more commonly used analog technologies here. Because no matter if you're using a real Fairchild or

Sidechain Input

Many dynamics processors allow you to send an external signal to the detector. This way, the processor can 'listen' to a signal other than the one it's manipulating. If you want to do this digitally, besides audio inputs and outputs, your plugin will also need a sidechain input. And the host in which you use the plugin has to support sidechaining, via an internal bus or aux-send. If that's the case, you can easily put together very complex sidechain constructions. For example, you can make a number of sound sources trigger the compression on another sound source, or vice versa. In EDM, this method is used a lot to make the downbeat push the rest of the instruments away. If you can find the right dosage for this effect, it can produce a very powerful but still dynamic mix (the turning point is when the offbeat starts to feel bigger than the downbeat).

Gates are at least as interesting to control externally as compressors. You can use them to make an instrument or sound effect closely follow the course of another element. Turning on the room mics on the drums only when the toms are played is a well-known example, but the possibilities are endless. Sidechain constructions are mainly interesting if you want to manipulate the feel of the rhythm or create room in your mix at selected moments. There are also special 'sidechain' plugins that can generate the typical bounce effect without the need for an actual sidechain construction. These are actually envelope generators that generate a certain volume progression synchronously with the beat. The advantage of this is that the volume progression (the shape of the fades, more or less) is often more precisely adjustable than on a compressor.

the digital version—if you know how it works and what sets it apart, you'll automatically use it more efficiently in your mix.

Vari-Mu

The vari-mu is the oldest type of compressor, dating back to the 1940s and originally developed as a broadcasting limiter. Famous classics are the Universal Audio 175, Altec 436, Gates STA-Level and of course the Fairchild 670. In a vari-mu compressor, the gain reduction element is a special type of tube, whose amplification (mu) can be influenced by varying the voltage difference (bias voltage) between the control grid and the tube's cathode. The wires of the control grid are distributed in such a way that the current between the anode and cathode of the tube can never be completely cut. This is why a vari-mu compressor has a maximum level reduction that's dependent on the kind of tube used (typically about 15 dB, but sometimes a lot more). As a result, there's a limit to the amount of compression, which contributes to the fact that many vari-mus still sound acceptable when their threshold is far exceeded (see Figure 17.26). This is an interesting feature, which is now also built into other compressors in the form of gain reduction limiting.

Due to the fact that the (DC) bias voltage in vari-mu compressors is combined with the (AC) audio signal, and varied in order to control the compression, subsonic pulses occur in the audio signal during these changes. Manufacturers try to minimize this effect by creating a balanced circuit (this way, they can cancel out the control signal), but they never completely pull it off. This is not the end of the world though, because the mild subsonic pulse that slips through in certain compressors can sometimes work very well to emphasize the rhythm of the music. For example, when the compressor is triggered by every bass and snare drum, the extra sub-lows will only be added in those places.

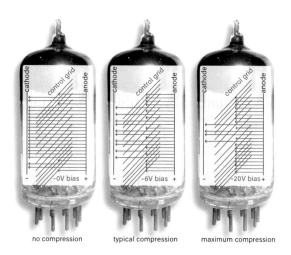

Figure 17.26 The control grid of a vari-mu tube is more coarse in the middle of the tube than at both ends. By varying the bias voltage on the control grid, the number of electrons that can flow from the anode to the cathode (and therefore the amount of amplification) can be controlled. However, the flow of electrons can never be completely blocked in the middle of the tube. This limits the maximum amount of compression that's possible.

The attack and release times of vari-mu systems are on the slow side (the fastest attack time is typically about 1 to 5 milliseconds, with some exceptions), which generally makes the vari-mu a relatively mild compressor. Therefore, it's less suitable as a peak limiter, but it's a great tool to gradually make a sound more compact. The fastest attack option on some modern vari-mus is often a 'close, but no cigar' intermediate option if you ask me: you never get the control of the transients you're hoping for with such a fast attack, and at the same time there's not enough of the transient passing through, so the whole thing doesn't breathe anymore. The sound becomes stifled and swampy, because the transients trigger a reaction that can't keep up with their own fast nature, in a way. The simple solution is to use a different type of compressor (FET, VCA, digital) for the peaks, and not take the vari-mu out of its natural habitat: just set the attack a bit slower! (The classic Altec 436 even had a fixed attack time of 50 ms.) Because all vari-mu compressors have tubes in their signal path, they often sound 'warm' due to the harmonic distortion these tubes create, and with more compression, this distortion increases. With extreme amounts of compression, the distortion can be more than 1 percent, but still sound acceptable. An advantage of this is that you need less compression to give a sound a certain 'presence,' so you can leave more of the dynamics intact.

Opto

The Teletronix LA-2A has been a household name since the early sixties, especially as a compressor for vocals. So what's so special about this design? Fundamentally, every compressor consists of a detector and an attenuator. In an optical (opto) compressor such as the LA-2A, the detector is a light that's connected to the audio. The attenuator is a light-dependent resistor (LDR) that's connected to ground in parallel with the audio. As the level of the incoming audio increases, the light shines brighter on the LDR. This causes its resistance to decrease, so more audio can leak to ground and the audio level is reduced (see Figure 17.27).

What's special about this design is the timing it provides, due to the fact that the light source and LDR take some time to get going. That's why the first designs—which had regular light bulbs—were quickly upgraded, as the time the bulb needed to start glowing made the compressor's attack time impracticably long. And after the system is activated, it also takes a while to get back to its normal state. This return time depends on the duration of the light that shines on the LDR, and the course of the return isn't linear. You can compare it to the plastic glow-in-the-dark stars on the ceiling of a child's bedroom: the longer the light is on before it's time for bed, the longer the stars will shine. Plus they don't become weaker linearly: right after the light is turned off they

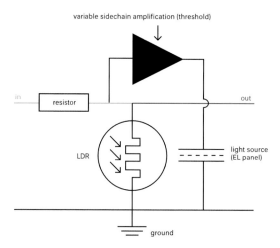

Figure 17.27 How an optical compressor works. The stronger the audio signal, the more light the EL panel emits, the lower the resistance of the LDR, and the more of the signal 'leaks' to the ground.

shortly shine brightly, after which they gradually become weaker for a relatively long time. In an opto compressor, this means that short peaks cause short release times, while a longer exceedance of the threshold means a longer release time. So you could say that the LDR has a memory.

Opto compressors don't lend themselves to extreme, aggressive compression. The time constants are simply too slow for this. Modern designs try to push these limits, but swiftness will never be their strong suit. The magic of an opto compressor is in its signal dependence. Besides the attack and release time, the threshold and ratio are also signal-dependent, as these can be different per frequency. The exact sonic effects of this are different for every light source–LDR combination. There are designs with light bulbs, LEDs and electroluminescent panels (such as the T4 cell in the LA-2A). Each of these designs has its own sound character and timing, but in general they respond very naturally to source material with long notes and without very sharp transients, such as vocals, strings and bass.

The signal dependency that you get 'for free' with an optical system is often added to other compressor technologies as well. By building timing-sensitive circuits into the detector, these designs can also have an auto-release function, but with a potentially much faster attack time than an optical compressor. And another advantage of these compressors is that you can choose if you want to use auto-release or not. With an opto compressor you don't have that choice, so for percussive source material the release will often be too slow.

Another unique side effect of the system is that the control signal isn't electrically connected to the audio path. There is only an optical connection, which prevents the audio signal from affecting the control signal, and vice versa. Many other designs do have an electrical connection, which can cause distortion. Optos don't have this distortion, and can therefore sound incredibly transparent. The warm sound they are associated with usually comes from the circuit around the LDR, which can consist of tubes and transformers.

Diode Bridge

Diode bridge compressors were developed in the early 1950s for German broadcasters, as an alternative to the high-maintenance and expensive vari-mu compressors. Though both of these operating principles have a lot of similarities, the diode bridge has its own distinctive sonic character. This character can be heard throughout a considerable part of pop history, thanks to the compressors developed by EMI for its own use. Later, the technology became almost synonymous with Neve compressors.

As the name suggests, a diode bridge compressor uses diodes as an attenuator. A diode only allows electrons to flow in one direction. However, this conduction won't start until a certain voltage difference between the input and output of the diode is reached, and it becomes easier as the voltage difference increases. In other words: diodes have an electrical resistance that's dependent on the voltage difference between the input and output poles. The higher this voltage difference, the lower the diode's resistance. So within a certain voltage range, a diode can be used as a variable resistor by connecting a control voltage (bias) to the diode. In a diode bridge compressor, this control voltage is dependent on the signal level of the audio. The resistance of the diodes—which in most cases are connected in a network (bridge) of four diodes—gets lower as the audio signal further exceeds the threshold. To make the diodes suitable as a compression element, they have to be used in a shunt (parallel) configuration: in a way, they form a variable 'leak' in the signal path, whose resistance decreases as the control voltage increases. The lower this resistance, the more audio 'leakage'—and thus compression—occurs (see Figure 17.28).

As with vari-mu compressors, the DC control voltage is combined with the AC audio signal: both are placed across the diodes. Therefore, in this design the circuit should be balanced as well in order to remove the control voltage from the audio after compression. What's special about diode bridge compressors is that the audio signal contributes significantly to the control voltage. This way, the audio causes distortion, but only when the diodes are conducting and compression is actually taking place. In neutral mode, nothing happens: the control voltage is absent, and the

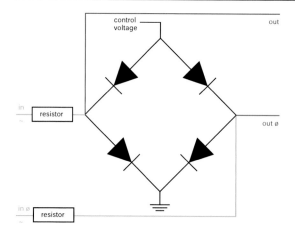

Figure 17.28 The circuit diagram of a diode bridge compressor. Two diodes are needed to form a variable resistor for (AC) audio signals. And because the circuit has a balanced arrangement, the bridge consists of four diodes.

audio signal alone doesn't cause enough voltage difference to make the diodes conduct. The sound of diode bridge compression resembles that of vari-mu, in the sense that it's smooth and fluid. But diode bridge compressors can intervene faster and more acutely, and their distortion has a very typical, slightly grainy, but warm sound.

FET and PWM

To this day, the Urei 1176 is one of the fastest-responding analog compressors. This has everything to do with the compression element it uses: a field-effect transistor (FET). This type of transistor lends itself well for use as a variable resistor, plus it has an exceptionally fast response. The resistance value is dependent on a control voltage: the higher the control voltage, the lower the resistance. In a parallel configuration, a FET can form a variable 'leak' to the ground, thus attenuating the audio signal (see Figure 17.29).

The drawback of FETs is that they usually can't handle line-level audio signals, as these signals affect (modulate) the FET's resistance during compression. To prevent large amounts of distortion, the input signal of a FET compressor must therefore be attenuated. Nevertheless, good designs can still achieve a workable dynamic range. The fact that FETs can respond faster than other technologies doesn't mean that a FET compressor responds quickly by definition. A FET can form the basis of a compressor with a highly versatile response, which can intervene

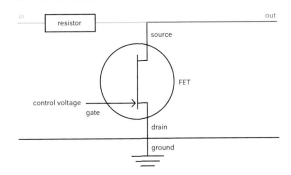

Figure 17.29 The operation of a FET compressor. By varying the control voltage on the gate of the FET, the amount of current that can leak from the source to the drain—and therefore to the ground—can be influenced.

as quickly and acutely as the 1176, but also as smoothly as an opto compressor.

The Empirical Labs Distressor is a well-known example of a FET compressor with a highly manipulable sidechain, which makes it a chameleon. Some manufacturers—such as EMT, Pye and MXR in the past, and nowadays Crane Song—go even further and use FETs in a completely different way: in a pulse width modulation (PWM) configuration.

PWM can be seen as a system with a 1-bit resolution (the possible values are one or minus one) and a very high sampling rate. It's like a switch that can be turned on and off very quickly. An audio signal is converted by the system into a series of pulses that represent the average energy of the audio signal. Technically, it roughly works as follows: the level of the audio signal is continuously compared with a high-frequency triangle wave. If the level is higher, the output value is one; if it's lower, the output value is minus one. The result is a series of pulses with a pulse width that's related to the original signal level. This series can then be converted back into an audio signal by making the pulses switch a voltage source on or off. The resulting signal is a square wave that only needs to be filtered to rid it of byproducts (the sharp corners), after which it's virtually identical to the source signal again.

The voltage source that's switched on and off during decoding can have a higher voltage than required for a one-on-one reconstruction of the source signal, thus amplifying the source signal. This is the—completely analog—principle on which the class-D amplifier is based (see Figure 17.30).

This system can also produce variable amplification by digitizing the series of pulses and manipulating them with an algorithm. In a PWM compressor, the audio is converted into a series of pulses, which are digitally manipulated based on the user settings in the sidechain.

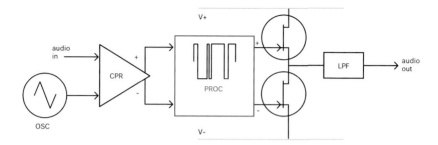

Figure 17.30 In a PWM compressor, a comparator compares the audio signal level with a high-frequency triangle wave. If it's higher, the positive output of the comparator produces a voltage; if it's lower, the negative output produces a voltage. This creates a series of pulses that can switch the two FETs on and off, so they alternately allow current from the positive and negative side of the voltage source to pass. This way, a square wave is constructed, which—after being filtered by a low-pass filter—is identical to the input signal, so no compression takes place. However, the series of pulses can also be (digitally) manipulated inside the red square, thus modifying the output signal.

Next, the sidechain signal is made analog again and sent to a set of FETs that reconstruct the manipulated signal. Because the FETs are only fully switched on or off during this reconstruction, the transition area in which the FETs might not respond linearly is bypassed. This creates an amplifying element that doesn't increase distortion with the amount of compression, and that has an extremely constant frequency response. The possibility of digitally manipulating the sidechain signal makes PWM compressors highly flexible in terms of configuration: all kinds of complex attack and release characters are possible. As a result, a PWM compressor sounds nothing like a stereotype FET compressor.

VCA

VCA (voltage-controlled amplifier) is a confusing name for a technology, since every analog compressor technically uses a voltage-controlled amplifier. However, VCA became synonymous with an amplifying element consisting of a network of transistors, invented by DBX founder David Blackmer. His first model 202 VCA used transistors in an oven, so the operating temperature of the transistors would be high, but constant. This way, he managed to bypass the temperature-dependent response of the transistors and achieve a constant control voltage-amplification ratio. The DBX 202 was the first VCA with a dynamic range and distortion character that could be used for audio, and also the first exponentially responding VCA: the ratio between the control signal and amplification is logarithmic. A two-fold increase of the control voltage causes an equal increase in amplification (in decibels). This ratio remains constant throughout the VCA's dynamic range. This is a significant difference from other technologies such as FET, diode bridge, vari-mu and opto, which all to some extent respond non-linearly to a control voltage.

Since the DBX 202, development hasn't stood still, and of all the attenuation elements found in modern-day compressors, VCAs have the largest dynamic range. Thanks to this quality, they are widely used in mixer automation, noise reduction systems and gates, among other applications. The famous SSL bus compressor even shares its VCA with the master fader of the mixing console it's built into! Today, a VCA can consist of discrete transistors—as in the old DBX 160—or have all the transistors integrated on a chip. For example, many modern compressors are based on a chip made by THAT Corporation. What's more, many of these compressors use the exact same chip. But they still all sound different, so apparently this difference has little to do with the VCA.

Because the VCA itself has an almost perfect response, the design of the sidechain largely determines the sound of the compressor. In that sense it resembles a digital compressor: a perfectly responding attenuation element is easily made, but the non-linear behavior and signal dependency—which make an optical element work so well on lead vocals, for example—don't come for free.

Compare and Choose

It should be clear by now that you can choose from a huge variety of sound characters when you're looking for a compressor. But within this vast range, how do you choose a compressor for a particular application? First, it's important to decide for yourself what you want to achieve with the compressor, and there are actually only three possibilities:

- amplify a good characteristic of a sound;
- attenuate an unpleasant characteristic of a sound;
- change as little as possible of the character of a sound and only reduce its dynamics.

A characteristic of a sound can of course be anything: a certain timbre, a rhythm, the transients or the decay, and so on. Once you have a clear image of what the compressor is supposed to do, it suddenly becomes very easy to make a choice. For example: you want to compress a sibilant vocal recording. This unpleasant characteristic is something you want to attenuate, or at least not exaggerate. Therefore, you won't choose a compressor that responds too much to the lows in the voice, as this would make it sound smaller and emphasize the high frequencies more.

It's better to choose a compressor with a warm sound and slower response times, or one that sounds progressively duller with greater amounts of gain reduction.

In order to work this way, you have to be familiar with the different sounds you can make. If you don't have this clearly in mind yet, the best way to learn is by setting up a session with a variety of instruments: drums, bass, strings, piano, synths, acoustic guitar and vocals. Next, you send this audio to as many buses as you have compressors, and try to systematically determine where the strengths and weaknesses of these compressors are with different source material.

Beyond Hardware

Although the ultimate goal for many compressor plugins is still to sound exactly like their analog forebears, there are some that make full use of the added possibilities of digital processing. With plugins, it's possible to use infinitely fast attack times, for example. Thanks to a buffer, a limiter can 'glimpse into the future' (look ahead) and see peaks coming in advance. As a result, digital limiters can respond faster and sound more transparent than their analog counterparts. But much more exotic manipulations are also possible. For instance, the Dynamic Spectrum Mapper (DSM) by Pro Audio DSP is an FFT compressor. Just like a spectrum analyzer, this compressor splits the signal into 1024 bands, each of which can have its own threshold setting. With a capture function, you can 'teach' the compressor a new threshold curve based on the input signal. More about spectral processing can be found in section 9.6.

The possibilities in the digital domain are endless, but still it's not all that simple to make a digital compressor sound good. Some positive side effects that you get for free in analog devices have to be designed and implemented from the ground up in the digital realm. This requires a lot of research, but in the end it's worth it. After all, the big advantage of software is that there are fewer restrictions: you can build the most complex detection circuits without having to spend a fortune on electronic components. This creates a lot of room for innovation, which allows new classics to be born. And that's a lot more interesting than imitating the old ones!

COMPRESSOR STEREOTYPES

Technology	Examples (analog)	Attack character	Release character	Special features
Vari-mu	Fairchild 670, Gates STA-Level, Universal Audio 175 Thermionic Culture Phoenix, Gyraf Gyratec X, Manley Vari-Mu, Retro Instruments STA-Level & 176.	Average: not extremely fast, but not distinctly slow either. Has a **rather mild** response due to the soft knee and limited amount of compression, but can react more sharply than an opto compressor.	Adjustable and often on the slow side. Fairchild is known for its (frequently imitated) triple release: a complex auto-release that can take several seconds to bring the gain reduction back down to zero. **'Solid and sluggish.'**	Thumb effect: the control voltage leaks into the audio path, causing a subsonic pulse at the onset of reduction. A balanced circuit corrects this in part. Because vari-mu by definition uses tubes in the audio path, the technology is mainly associated with 'warmth.' The amount of compression is limited, so this type of compressor rarely goes 'over the top': it's ideal for giving bass sounds more overtones without flattening their dynamics too much.
Opto	Teletronix LA-2A & LA-3A, Pendulum OCL-2, Manley ELOP, Tube-Tech CL 1B, Shadow Hills Optograph, ADL 1500, Inward Connections TSL-4, Summit Audio TLA-100A.	On the slow side (sometimes adjustable from slow to even slower), signal-dependent and therefore **open/natural**-sounding. Modern versions (Pendulum/Buzz/Inward Connections) can respond a bit faster.	First fast, then slow, but generally on the slow side. 'Memory' after big or long-lasting reduction. Slower for low frequencies with a lot of energy, and—though sometimes adjustable in terms of time—always signal-dependent. **'Fluid.'**	An optical element with good performance in the audio band can be used to build highly transparent and musically responding compressors. There is no leak of the control signal into the audio path, and therefore little distortion (the LA-2A's coloration is mainly caused by the tube stage, not by the opto cell). Optos can often handle low frequencies quite well: they don't 'choke' on the lows as easily as other compressors do, thanks to their signal dependency.
FET	Universal Audio 1176, Daking FET Compressor, Purp e Audio MC77, Chandler German um Compressor, Emp rical Labs Distressor.	**Extremely fast.** The attack is usually too fast to introduce punch the way VCA or diode bridge compressors can, though there are models with longer attack times on board.	Adjustable from extremely fast (there are even FET brickwall limiters) to slow. The reason for using a FET compressor is often the fast release, which can bring out the space in a recording and create a lot of density. **'Presence, aggression & energy.'**	The distortion in FET compressors is on the high side, although this can be compensated through feedback. Often, the noise floor is on the high side as well, but for pop music applications this is rarely a problem. The weakness of some FET compressors is that they can get 'jammed' by large concentrations of low frequencies, due to their short attack time. This type of compressor is very suitable for bringing out the room sound in drum recordings, and for making vocals extremely 'in your face.'
VCA	DBX 160, SSL G-series Bus Compressor, Smart Research C2, API 2500, Vertigo Sound VSC-2, Elysia Alp1a, Mpressor & Xpressor.	Variable from very fast to slow. In feedforward configuration, the VCA is known for its hard knee response. In combination with a slightly longer attack, it can create **punch** on a mix or drum bus.	Variable from fast to slow. Simple designs can sound 'static' and tend to not follow the music enough, but there are more complex auto-release circuits that can fix this. **'Clean, precise & punchy.'**	VCAs have a large dynamic range and can produce a lot of reduction without distorting. Modern VCA compressors can be very clean-sounding. Still, many VCA compressors tend to get 'jammed' (high frequencies become dull and dynamics too unnatural) when very large amounts of gain reduction are required. Opto and vari-mu generally sound better when they are really pushed. Its controllable (non-signal-dependent) character makes VCA suitable for rhythmic manipulations.
Diode bridge	Neve 2254 & 33609, EMI TG Limiter, Chandler TG1 Limiter, Zener Limiter & LTD-2, Vintagedesign C.1.	Not as fast as FET or VCA, faster than opto, similar to vari-mu. The difference with vari-mu is in the way it intervenes: since it's not as limited, it can be **quite fierce.**	Variable from fast to slow. The sound can be very **energetic** and extreme, but at the same time also **warm and mild** as a result of the distortion.	Diode bridge functions as a kind of solid-state vari-mu circuit: the bridge is biased, and adjusting this bias changes the resistance. The audio is combined with the control voltage, and affects this voltage as more compression takes place. As a result the distortion is very high, but quite full and creamy-sounding. This type of compression works well on bright sources such as overheads and women's voices; it makes them sound fuller.
PWM	EMT 156, Crane Song Trakker & STC-8, D.W. Fea rn VT-7, Summit Audio DCL-200, Pye, MXR Dual Limiter, Dave Hill Designs Titan.	Adjustable from extremely fast to slow: **versatile.**	Adjustable from extremely fast to slow: **versatile, but always on the clean side.**	A FET is used in a class-D configuration, controlled by a PWM circuit. Distortion is low and independent of the compression. In terms of compression sound character, this type of compressor can be a true chameleon, but the coloration that a tube, FET or diode would cause is missing.

Figure 17.31 The attack and release character can be quite dependent on how the sidechain is structured. A compressor can never be faster than the maximum capacity of its technology, but a FET compressor can also respond slowly, for example. However, the table shows the stereotype (a fast FET compressor), which often constitutes the reason for choosing a particular type of compressor.

Recommended Literature

A lot has been written about the subject of this book, and online you'll find countless tutorials that can teach you all sorts of techniques. But sources that provide insight into why you would do something (instead of just explaining how) are pretty rare. Therefore, I can highly recommend these books to you:

Michael Paul Stavrou, *Mixing with Your Mind*
Larry Crane (Ed.), *Tape Op: The Book About Creative Music Recording* (2 vols)
Mixerman, *Zen and the Art of Mixing*

And a bit more on the technical side, but essential reading for every engineer:

Bob Katz, *Mastering Audio*
Floyd Toole, *Sound Reproduction*

Besides reading books by and for engineers, I sometimes find it helpful to look at our line of work from a distance, and see how it relates to how others make and experience music. These books are refreshing and inspiring in that regard:

David Byrne, *How Music Works*
Greg Milner, *Perfecting Sound Forever*
Daniel J. Levitin, *This Is Your Brain on Music*

Index

Taylor & Francis eBooks

www.taylorfrancis.com

A single destination for eBooks from Taylor & Francis
with increased functionality and an improved user
experience to meet the needs of our customers.

90,000+ eBooks of award-winning academic content in
Humanities, Social Science, Science, Technology, Engineering,
and Medical written by a global network of editors and authors.

TAYLOR & FRANCIS EBOOKS OFFERS:

A streamlined
experience for
our library
customers

A single point
of discovery
for all of our
eBook content

Improved
search and
discovery of
content at both
book and
chapter level

REQUEST A FREE TRIAL
support@taylorfrancis.com